50
IDEAS YOU REALLY NEED TO KNOW
CHEMISTRY

HAYLEY BIRCH

Quercus

Contents

Introduction

Chemistry is often looked upon as the underdog of the sciences. I was speaking to a chemist only the other day who told me she was fed up with her subject being viewed as 'just a bunch of people mucking around with smelly things in labs'. Somehow, chemistry is thought of as less relevant than biology and less interesting than physics.

So as an author of a chemistry book, my challenge is to help you get over this image problem and root for the underdog. Because – and not many people know this – chemistry is actually the best science.

Chemistry is at the heart of pretty much everything. Its building blocks – the atoms, molecules, compounds and mixtures – make up every ounce of matter on this planet. Its reactions are responsible for supporting life and creating everything that life depends on. Its products chart the progress of our modern existence – from beer to Lycra hotpants.

The reason chemistry has an image problem, I think, is that rather than focussing on the relevant, interesting stuff, we get bogged down trying to learn a set of rules for how chemistry works, formulae for molecular structures, recipes for reactions and so on. And while chemists may argue that these rules and recipes are important, most will agree that they're not particularly exciting.

So we won't be dealing very much with rules in this book. You can look them up somewhere else if you like. I've tried to keep the focus on what I think is relevant and interesting about chemistry. And along the way, I've tried to channel the spirit of my chemistry teacher Mr Smailes, who showed me how to make soap and nylon, and wore some really excellent ties.

01 Atoms

Atoms are the building blocks of chemistry, and of our Universe. They make up the elements, the planets, the stars and you. Understanding atoms, what they're made of and how they interact with each other, can explain almost everything that happens in chemical reactions in the lab, and in nature.

Bill Bryson famously wrote that each one of us might be carrying up to a billion atoms that once belonged to William Shakespeare. 'Wow,' you may well think, 'That's a lot of dead Shakespeare atoms.' Well, it is and it isn't. On the one hand, a billion (1,000,000,000) is about the number of seconds that each of us will have lived on our 33rd birthday. On the other hand, a billion is the number of grains of salt that would fill an ordinary bath, and less than one billionth of one billionth of the number of atoms in your entire body. This goes someway to explaining how small an atom is – there are over a billion times a billion times a billion just in you – and suggests that you don't even have enough dead Shakespeare atoms to make up one brain cell.

LIFE'S A PEACH

Atoms are so tiny that, until recently, it was impossible to see them. That fact has changed with the development of superhigh resolution microscopes, to the point where, in 2012, Australian scientists were able to take a photograph of the shadow cast by a single atom . But chemists didn't always have to see them to understand that, at some fundamental level, atoms could explain

TIMELINE

*c.*400BC	1803	1904	1911
Greek philosopher Democritus refers to indivisible atom-like particles	John Dalton proposes atomic theory	Joseph John Thomson's 'plum pudding model' of the atom	Ernest Rutherford describes the atomic nucleus

most of what goes on in the lab, and in life. Much of chemistry is down to the activities of even tinier, subatomic particles called electrons, which make up the atom's outer layers.

If you could hold an atom in your hand like a peach, the stone in the middle would be what is called the nucleus, containing the protons and neutrons, and the juicy flesh

Atomic theory and chemical reactions

In 1803, the English chemist John Dalton gave a lecture in which he proposed a theory of matter based on indestructible particles called atoms. He said, in essence, that different elements are made of different atoms, which can combine to make compounds, and that chemical reactions involve a rearrangement of these atoms.

would be made up of electrons. In fact, if your peach was really like an atom, most of it would be flesh and its stone would be so small you could swallow it without noticing – that's how much of the atom is taken up by electrons. But that core is what stops the atom drifting apart. It contains the protons, positively charged particles that hold just enough attraction for the negatively charged electrons to stop them flying off in all directions.

WHY IS AN OXYGEN ATOM AN OXYGEN ATOM?

Not all atoms are the same. You may already have realized that an atom doesn't share that many similarities with a peach, but let's extend the fruit analogy further. Atoms come in many different varieties or flavours. If our peach was an atom of oxygen, then a plum might be, say, an atom of carbon. Both little balls of electrons surrounding a proton pip, but with completely different characteristics. Oxygen atoms float around in pairs (O_2) while carbon sticks together en masse to make hard substances like diamond and pencil lead (C). What makes them different elements (see page 8) is their respective numbers of protons. Oxygen, with eight protons has two more than carbon. Really large, heavy elements like seaborgium and nobelium have more than one hundred protons in each of their atomic cores. When

1989

IBM researchers manipulate individual atoms to spell 'IBM'

2012

Discovery of the Higgs boson adds to standard model of the atom

Splitting the atom

J.J. Thomson's early 'plum pudding' model of the atom viewed it as a doughy positively charged 'pudding' with negatively charged 'plums' (electrons) distributed evenly throughout. That model has changed: we now know that protons and other subatomic particles called neutrons form the tiny, dense centre of the atom, and the electrons a cloud around them. We also know that protons and neutrons contain even smaller particles called quarks. Chemists don't generally dwell on these smaller particles – they are the concern of physicists, who smash up atoms in particle accelerators to find them. But it is important to remember that science's model of the atom, and of how matter fits together in our Universe, is still evolving. The discovery of the Higgs boson in 2012, for example, confirmed the existence of a particle that physicists had already included in their model and used to make predictions about other particles. However, there's still work to do to determine if it's the same type of Higgs boson they were looking for.

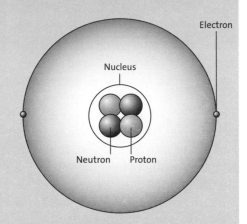

The incredibly dense nucleus of an atom contains positively charged protons and neutral neutrons, orbited by negatively charged electrons.

there are this many positive charges crammed into the vanishingly small space of the nucleus, each repelling the other, the equilibrium is easily upset and heavy elements are unstable as a result.

Usually, an atom, whatever its flavour, will have the same number of electrons as it has protons in its core. If an electron goes missing, or the atom collects an extra one, the positive and negative charges no longer balance each other out and the atom becomes what chemists call an 'ion' – a charged atom or molecule. Ions are important because their charges help stick together all sorts of substances, such as the sodium chloride of table salt and the calcium carbonate of limescale.

THE BUILDING BLOCKS OF LIFE

Besides making up kitchen cupboard ingredients, atoms form everything that crawls or breathes or puts down roots, building stunningly complex molecules like DNA and the proteins that form our muscles, bones and hair. They do this by bonding (see page 20) with other atoms. What's interesting about all life on Earth, however, is that despite its tremendous diversity, without exception it contains one particular flavour of atom: carbon.

From bacteria clinging to life around smoking hot vents in the deepest, darkest parts of the ocean to birds soaring in the sky above, there is not a living thing on the planet that doesn't share that element, carbon, in common. But because we have not yet discovered life elsewhere, we can't tell whether it was random chance that life evolved this way, or whether life could thrive using other types of atoms. Science-fiction fans will be well acquainted with alternative biologies – silicon-based beings have appeared in *Star Trek* and *Star Wars* as alien life forms.

ATOM BY ATOM

Progress in the area of nanotechnology (see page 180), which promises everything from more efficient solar panels to drugs that seek and destroy cancer cells, has brought the world of the atom into sharper focus. The tools of nanotechnology operate at a scale of one billionth of a metre – still bigger than an atom, but at this scale it is possible to think about manipulating atoms and molecules individually. In 2013, IBM researchers made the world's smallest stop-motion animation, featuring a boy playing with a ball. Both the boy and the ball were made from copper atoms, all visible individually in the movie. Finally, science is starting to work at a scale that matches the chemist's view of our world.

> **THE BEAUTY OF A LIVING THING IS NOT THE ATOMS THAT GO INTO IT, BUT THE WAY THOSE ATOMS ARE PUT TOGETHER.**
> Carl Sagan

The condensed idea
Building blocks

02 Elements

Chemists go to great lengths to discover new elements, the most basic chemical substances. The Periodic Table gives us a way to order their discoveries, but it's not just a catalogue. Patterns in the Periodic Table provide clues about the nature of each element and how they might behave when they encounter other elements.

The 17th-century alchemist Hennig Brand was a gold digger. After getting married, he left his job as an army officer and used his wife's money to fund a search for the Philosopher's Stone – a mystical substance or mineral that alchemists had been seeking for centuries. Legend had it that the Stone could 'transmute' common metals like iron and lead into gold. When his first wife died, Brand found another and continued his search in much the same fashion. Apparently, it had occurred to him that the Philosopher's Stone could be synthesized from bodily fluids, so Brand duly acquired no less than 1500 gallons of human urine from which to extract it. Finally, in 1669, he made an astounding discovery, but it wasn't the Stone. Through his experiments, which involved boiling and separating the urine, Brand had unwittingly become the first person to discover an element using chemical means.

Brand had produced a compound containing phosphorus, which he referred to as 'cold fire' because it glowed in the dark. But it took until the 1770s for phosphorus to be recognized as a new element. By this time, elements were being discovered left, right and centre, with chemists isolating oxygen, nitrogen, chlorine and manganese all within the space of one decade.

TIMELINE

1669	1869	1913
First element – phosphorus – discovered by chemical means	Mendeleev publishes the first incarnation of his Periodic Table	Henry Moseley defines elements by their atomic number

In 1869, two centuries after Brand's discovery, the Russian chemist Dmitri Mendeleev devised the Periodic Table and phosphorus assumed its rightful place therein, between silicon and sulfur.

WHAT'S AN ELEMENT?

For much of history, 'the elements' were considered to be fire, air, water and earth. A mysterious fifth, aether, was added to account for the stars, since they could not, as the philosopher Aristotle argued, be made from any of the earthly elements. The word 'element' comes from a Latin word (*elementum*) meaning 'first principle' or 'most basic form' – not a bad description, but it does leave us wondering about the difference between elements and atoms.

Decoding the Periodic Table

In the Periodic Table (see pages 204-5) elements are represented by letters. Some are obvious abbreviations, such as Si for silicon, while others, such as W for tungsten, seem to make no sense – these are often references to archaic names. The number above the letter is the mass number – the number of nucleons (protons and neutrons) in the nucleus of an element. The subscripted number is its number of protons (atomic number).

The difference is simple. Elements are substances, in any quantity; atoms are fundamental units. A solid lump of Brand's phosphorus – incidentally, a toxic chemical and a component of nerve gas – is a collection of atoms of one particular element. Curiously though, not every lump of phosphorus will look the same, because its atoms can be arranged in different ways, changing the internal structure but also the outward appearance. Depending on how the atoms are arranged in phosphorus, it can look white, black, red or violet. These different varieties also behave differently, for instance, melting at wildly different temperatures. White phosphorus, for example, melts under the Sun on a very hot day, while black phosphorus would need to be heated in a roaring furnace at over 600 °C. Yet both are made from the very same atoms containing 15 protons and 15 electrons.

1937

The first artificially produced element – technetium

2000

Russian scientists make the superheavy element livermorium

2010

Discovery of element with atomic number 117 ('ununseptium') announced

PATTERNS IN THE PERIODIC TABLE

To the untrained eye, the Periodic Table (see pages 204–5) has the appearance of a slightly unorthodox game of Tetris, where – depending on the version you look at – some of the blocks have not quite dropped to the bottom. It looks like it needs a good tidy-up. Actually, it's a well-ordered mess and any chemist will quickly be able to find what he or she is looking for among the apparent disarray. This is because Mendeleev's cunning design contains hidden patterns that link together elements according to their atomic structures and chemical behaviours.

Along the table's rows, from left to right, the elements are arranged in order of atomic number – the number of protons that each element has in its core. But the genius of Mendeleev's invention was discerning when the properties of the elements began to repeat and a row should be turned. It is from the columns, therefore, that some of the more subtle insights are gleaned. Take the column on the far right, which runs from helium to radon. These are the noble gases, all colourless gases under normal conditions and all particularly lazy when it comes to being involved in any kind of chemical reaction. Neon, for instance, is so unreactive that it cannot be convinced to enter into a compound with any other element. The reason for this is related to its electrons. Within any atom, the electrons are arranged in concentric layers, or shells, which can only be occupied by a certain number of electrons. Once a shell is full, further electrons must start to fill a newer, outer layer. Since the number of electrons in any given element increases with increasing atomic number, each element has a different electron configuration. The key feature of the noble gases is that all of their outermost shells are full. This full structure is very stable, meaning the electrons are difficult to prod into action.

> **THE WORLD OF CHEMICAL REACTIONS IS LIKE A STAGE ... THE ACTORS ON IT ARE THE ELEMENTS.**
>
> Clemens Alexander Winkler, discoverer of the element germanium

We can recognize many other patterns in the Periodic Table. It takes more effort (energy) to prise an electron away from an atom of each element as you move from left to right, towards the noble gases, and from bottom to top.

The middle of the table is occupied mostly by metals, which become more metallic the closer you edge to the far left corner. Chemists use their understanding of these patterns to predict how elements will behave in reactions.

SUPERHEAVYWEIGHTS

One of the few things that chemistry shares in common with boxing is that both have their superheavyweights. While the flyweights float at the top of the Periodic Table – with atoms of hydrogen and helium carrying just three protons between them – those on the bottom rows have sunk by virtue of their heavy atomic loads. The table has grown over many years to incorporate new discoveries and heavier elements. But at number 92, the radioactive element uranium is really the last element to be found in nature. Although the natural decay of uranium yields plutonium, the quantities are vanishingly small. Plutonium was discovered in a nuclear reactor and other superheavyweights are made by smashing together atoms in particle accelerators. The hunt isn't over yet but it's certainly become a lot more complicated than boiling up bodily fluids.

The hunt for the heaviest superheavy

No one likes a cheat, but you'll find them in every profession and science is no exception. In 1999, scientists at the Lawrence Berkeley Laboratory in California had published a scientific paper celebrating their discovery of superheavy elements 116 (livermorium) and 118 (ununoctium) But something wasn't making sense. Having read the paper, other scientists had tried to repeat the experiments, but no matter what they did they couldn't seem to conjure a single atom of 116. It turned out one of the 'discoverers' had fabricated the data, leaving a US government agency to make an embarrassing climbdown from statements about the world-class science it was funding. The paper was pulled and the plaudits for discovering livermorium went to a Russian group a year later. The scientist who faked the original data was fired. Such is the prestige associated with discovering a new element these days that scientists are willing to stake their entire careers on it.

The condensed idea
The simplest substances

03 Isotopes

Isotopes aren't just deadly substances used to make bombs and poison people. The concept of an isotope is one that encompasses many chemical elements that have a slightly altered quota of subatomic particles. Isotopes are present in the air we breathe and the water we drink. You can even use them (perfectly safely) to make ice sink.

Ice floats. Except when it doesn't. Just as all atoms of a single element are the same, except when they are different. If we take the simplest element, hydrogen, we can agree that all atoms of this element have one proton and one electron. You couldn't call a hydrogen atom a hydrogen atom unless it had only one proton in its nucleus. But what if the single proton was joined by a neutron? Would it still be hydrogen?

Neutrons were the missing piece of the puzzle that eluded chemists and physicists until the 1930s (see The missing neutrons, opposite). These neutral particles make no difference at all to the overall balance of charge in an atom, but radically alter its mass. The difference between one and two neutrons in the core of a hydrogen atom is enough to make ice sink.

HEAVY WATER

Packing an extra neutron into a hydrogen atom makes a big difference – for these flyweight atoms, it's double the quota of nucleons. The resulting 'heavy hydrogen' is called deuterium (D or ^{2}H) and, just as regular hydrogen atoms do, deuterium atoms hook up with oxygen to make water. Of course, they don't make regular water (H_2O). They make water with extra neutrons

TIMELINE

1500s	1896	1920
Alchemists try to 'transmutate' substances into precious metals	First use of radiation in cancer treatment	Early description of 'neutral doublets' (neutrons) by Ernest Rutherford

The missing neutrons

The discovery of neutrons by physicist James Chadwick – who went on to work on the atomic bomb – solved a niggling problem with the weights of the elements. For years, it had been apparent that atoms of each element were heavier than they should be. As far as Chadwick was concerned, atomic nuclei couldn't possibly weigh as much as they did if they only contained protons. It was like the elements had turned up for their summer holidays with their baggage full of bricks. Only no one could find the bricks. Chadwick had become convinced by his supervisor Ernest Rutherford that atoms were smuggling subatomic particles. Rutherford described these so-called neutral doublets or neutrons in 1920. But it took Chadwick until 1932 to find the concrete evidence to back up the theory. He found that by bombarding the silvery metal beryllium with radiation from polonium, he could get it to emit neutrally charged subatomic particles – neutrons.

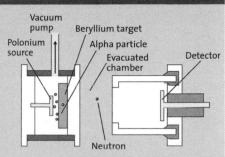

The reaction that knocks neutrons (n) from the beryllium target is: $^4_2He + ^9_4Be \rightarrow ^1_0n + ^{12}_6c$

in it: 'heavy water' (D_2O), or to give it its proper name, deuterium oxide. Take heavy water – easily purchased online – and freeze it in an ice-cube tray. Plop a cube into a glass of ordinary water and, bingo, it sinks! For comparison, you can add an ordinary ice cube and marvel at the difference that one subatomic particle per atom makes.

In nature, about one in every 6,400 hydrogen atoms have an extra neutron. There is, though, a third type – or isotope – of hydrogen, and this one is much rarer and rather less safe to handle at home. Tritium is an isotope of hydrogen in which each atom contains one proton and two neutrons. Tritium is unstable, however, and like other radioactive elements it undergoes radioactive decay. It is used in the mechanism that triggers hydrogen bombs.

1932

James Chadwick discovers the neutron

1960

Nobel Prize for Chemistry awarded to Willard Libby for radiocarbon dating using carbon-14

2006

Alexander Litvinenko dies of radioactive polonium poisoning

RADIOACTIVITY

Often the word 'isotope' is preceded by the word 'radioactive', so there might be a tendency to assume that all isotopes are radioactive. They are not. As we have just seen, it is perfectly possible to have an isotope of hydrogen that is non-radioactive – in other words, a stable isotope. Likewise, there are stable isotopes of carbon, oxygen and other elements in nature.

Unstable, radioactive isotopes decay, meaning that their atoms disintegrate, shedding matter from their core in the form of protons, neutrons and electrons (see Types of radiation, below). The result is that their atomic number changes and they can become different elements altogether. This would have seemed like magic to 16th- and 17th-century alchemists who were obsessed with finding ways of changing one element into another (the other, ideally, being gold).

All radioactive elements decay at different rates. Carbon-14, a form of carbon with 14 neutrons in its nucleus instead of the usual 12, is safe to use without special precautions. If you were to measure out a gram of carbon-14 and leave it on a window ledge, you would be waiting a long time for its atoms to decay. It would take 5,700 years for around half of the carbon atoms in your sample to disintegrate. This measure of time, or decay rate, is called a half-life. By contrast, polonium-214 has a half-life of less than one thousandth of a second, meaning that in some crazy parallel world where you would be allowed to measure out a gram of radioactive polonium, you wouldn't even have a chance to get it to your window ledge before all it of had decayed dangerously.

The former Russian spy Alexander Litvinenko and, potentially, Palestinian leader Yasser Arafat were killed with a more stable isotope of polonium, which decays over days rather than seconds, albeit

Types of radiation

Alpha radiation consisting of two protons and two neutrons is equivalent to a nucleus of atomic helium. It is weak and can be stopped by a sheet of paper. Beta radiation is fast-moving electrons and penetrates skin. Gamma radiation is electromagnetic energy, like light, and can only be stopped by a thickness of lead. The effects of gamma radiation are very damaging and high-powered gamma rays are used to destroy cancerous tumours.

fatally. In the human body, the radiation released by disintegrating polonium-210 nuclei rips through cells, and causes pain, sickness and immune system shutdown as it does so. In investigations of these cases, scientists looked for the products of polonium decay, because the polonium-210 itself was no longer present.

BACK TO THE FUTURE

Radioactive isotopes can be deadly, but they can also help us understand our past. The carbon-14 we left slowly decaying on your window ledge has a couple of common scientific uses – one is radiocarbon dating of fossils, the other is learning about past climates. Because we have a good idea of how long radioactive isotopes take to decay, scientists are able to work out the age of artefacts, dead animals, or ancient atmospheres preserved in ice, by analysing levels of different isotopes. Any animal will breathe in small amounts of naturally occurring carbon-14 – in carbon dioxide – during its lifetime. This stops as soon as animals die and the carbon-14 in them starts to decay. Because scientists know that carbon-14 has a half-life of 5,700 years, they can work out when fossilized animals died.

When ice cores are drilled out from ice caps or glaciers that have been frozen for thousands of years, they provide a ready-made timeline of atmospheric change based on the isotopes they contain. These insights into our planet's past may help us to predict what will happen to our planet in the future, as carbon dioxide levels continue to change.

> **SELDOM HAS A SINGLE DISCOVERY IN CHEMISTRY HAD SUCH AN IMPACT ON THE THINKING IN SO MANY FIELDS OF HUMAN ENDEAVOUR.**
>
> Professor A. Westgren, presenting the Nobel Prize for Chemistry for radiocarbon dating to Willard Libby

The condensed idea
What a difference a neutron makes

04 Compounds

In chemistry, there are substances that contain only one element and those that contain more than one – compounds. It's when elements are put together that the extraordinary diversity of chemistry becomes apparent. It is hard to estimate how many chemical compounds there are, and with new ones being synthesized every year, they have a multitude of uses.

Occasionally in science, someone makes a discovery that contradicts what everyone believed was a fundamental law. For a while, people scratch their heads and wonder whether there was some mistake or whether the data was fudged. Then, when the evidence finally becomes irrefutable, the textbooks have to be rewritten and a whole new direction of scientific research opens up. Such was the case when Neil Bartlett discovered a new compound in 1962.

Working late on a Friday evening, Bartlett was alone in his lab when he made the discovery. He had allowed two gases – xenon and platinum hexafluoride – to mingle and produced a yellow solid. Bartlett, it turned out, had made a compound of xenon. Hardly surprising, you might think, but at the time most of the scientific community believed that xenon, like the other noble gases (see page 10), was completely unreactive and incapable of forming compounds. The new substance was named xenon hexafluoroplatinate and Bartlett's work soon convinced other scientists to start looking for other noble gas compounds. Over the next few decades, at least another 100 were found. Compounds containing noble elements have since been used to make anti-tumour agents and in laser eye surgery.

TIMELINE

1718	EARLY 1800s	1808
'Affinity table' showing how substances combine developed by Étienne François Geoffroy	Claude-Louis Berthollet and Joseph-Louis Proust debate the proportions in which elements combine	Chemical atomic theory by John Dalton confirms elements combine in fixed proportions

PARTNERING UP

Bartlett's compound may have been a turn-up for the books, but his story is not just a neat example of a scientific discovery overturning some widely held 'truth'. It is also a reminder of the fact that elements (especially unreactive ones) aren't all that useful on their own. To be sure, there are applications – neon lights, carbon nanotubes and xenon anaesthesia, to name just a few – but it's only by trying out new, and sometimes very complex, combinations of elements that chemists can make life-saving medicines and cutting-edge materials.

It takes one element partnering up with another, and maybe another and another, to create the useful compounds that form the basis of nearly all modern products, from fuels, fabrics and fertilisers, to dyes, drugs and detergents. There is hardly anything in your house that isn't made of compounds – unless, like the carbon in pencil lead, it's made of a simple chemical element. Even the things that grew or formed by themselves, like wood and water, are compounds. In fact, they're probably even more complicated..

Compounds or molecules?

All molecules contain more than one atom. Those atoms may be atoms of the same element, as in O_2, or atoms of different elements, as in CO_2. But of O_2 and CO_2, only CO_2 is a compound, because it contains atoms of different elements chemically bound together. So not all molecules are compounds – but are all compounds molecules? What makes matters confusing is ions (see Ions, page 19). Compounds whose atoms form charged ions don't really form molecules in the traditional sense. In salt, for example, a bunch of sodium ions (Na^+) are bonded to a bunch of chlorine ions (Cl^-) in a large, well-ordered and ever-repeating crystal structure. So there are not really independent 'molecules' of sodium chloride in the strictest sense. Here, the chemical formula, NaCl, shows the ratio of sodium ions to chloride ions, rather than referring to an isolated molecule. On the other hand, chemists will happily talk loosely about 'molecules of sodium chloride' (NaCl).

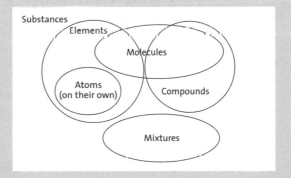

Substances — Elements, Molecules, Atoms (on their own), Compounds, Mixtures

1833
'Ions' defined by Michael Faraday and William Whewell

1962
Neil Bartlett shows noble gases can form compounds

2005
Chemical space estimates for 11-atom compounds of C, N, O and F

> **I TRIED TO FIND SOMEONE WITH WHOM TO SHARE THE EXCITING FINDING, BUT IT APPEARED THAT EVERYONE HAD LEFT FOR DINNER!**
>
> Neil Bartlett

COMPOUNDS AND MIXTURES

There are, however, some important distinctions we need to make when talking about compounds. Compounds are chemical substances that contain two or more elements. But just sticking two, or ten, elements in the same room as each other doesn't make them a compound. The atoms of those elements have to partner up – they have to form chemical bonds (see page 20). Without chemical bonding, all you have is a sort of cocktail party mingle bunch involving atoms of different elements – what chemists call a mixture. Atoms of some elements also partner with their own kind, such as oxygen in the air, which exists mostly as O_2, an oxygen doublet. The two oxygen atoms make an oxygen molecule. This oxygen molecule is not a compound either, because it only contains one type of element.

Compounds, then, are substances that contain more than one type of chemical element. Water is a compound, because it contains two chemical elements – hydrogen and oxygen. It's also a molecule, because it contains more than one atom. Most modern materials and commercial products are compounds that are composed of molecules too. But not all molecules are compounds, and it's debatable whether all compounds are molecules (see Compounds or molecules?, page 17).

POLYMERS

Some compounds are compounds within compounds – they are made up of basic units that are repeated many times, producing a beads-on-a-string effect. Such compounds are called polymers. Some of them you will recognize as polymers by their names – the polythene of your shopping bags, the polyvinyl chloride (PVC) of 'vinyl' LP records and the polystyrene of your takeaway boxes are all dead giveaways. Less obviously, nylon and silk, and the DNA inside your cells and the proteins in your muscles are also polymers. The repeating unit in all polymers, natural or man-made, is called a monomer. Stick the monomers together and you get a polymer. In the case of nylon, this makes for an impressive demonstration that is done in beakers in school chemistry labs everywhere – you can literally pull a length of nylon 'rope' from the beaker and wind it straight onto a bobbin, like a piece of thread.

BIOPOLYMERS

Biopolymers like DNA (see page 140) are so complex that it has taken millions of years of evolution for nature to perfect the art of making them. The monomers, or the 'compounds within the compound',

Ions

When an atom gains or loses a negatively charged electron, that change in the balance of charge causes the atom as a whole to become charged. That charged atom is called an ion. The same thing can happen with molecules, which form 'polyatomic' ions – a nitrate ion (NO_3^-) or a silicate ion (SiO_4^{4-}), for instance. Ionic bonding of oppositely charged ions is an important way of sticking substances together.

are nucleic acids, pretty complex chemicals in their own right. Linked together, they form long polymer strings that make up our DNA code. To link together DNA monomers, biology uses a special enzyme to add the individual beads to the string. It's incredible to think that evolution has found a way to make such complex compounds inside our own bodies.

Just how many compounds are out there? The honest answer is we don't know. In 2005, Swiss scientists tried to work out how many compounds containing just carbon, nitrogen, oxygen or fluorine would actually be stable. They reckoned nearly 14 billion, but that was only including compounds containing up to eleven atoms. The 'chemical universe' – as they called it – is truly vast.

The condensed idea
Chemical combinations

05 Sticking it all together

How does salt stick together? Why does water boil at 100 degrees centigrade? And most importantly, why is a lump of metal like a hippie commune? All of these questions and more are answered by paying close attention to the tiny negatively charged electrons that flit between and around atoms.

Atoms stick together. What would happen if they didn't? Well, for a start, the Universe would be a complete mess. Without the bonds and forces that holds materials together, nothing would exist as we know it. All the atoms that make up your body, and pigeons, flies, televisions, cornflakes, the Sun and Earth would swim around in a vast, near-infinite sea of atoms. So how do atoms become attached to each other?

NEGATIVE THINKING

In one way or another, atoms, within their molecules and compounds, are stuck together by their electrons – the tiny, subatomic particles that form a cloud of negative charge around an atom's positively charged nucleus. They order themselves into layers, or shells around the atomic nucleus and, since each element has a different number of electrons, each element has a different numbers of electrons in its outermost layer. The fact that an atom of sodium has an electron cloud that looks slightly different to the electron cloud of an atom of chlorine has some interesting effects, however. In fact, it's the reason they can stick to each other. Sodium easily loses the one electron in its outer shell. The loss of negative charge makes it positive

TIMELINE

1819	1873
Jöns Berzelius suggests chemical bonds are due to electrostatic attractions	Johannes Diderik van der Waals writes equation accounting for intermolecular forces in gases and liquids

(Na+). Meanwhile, chlorine easily gains a negatively charged electron to fill up its outer shell, becoming negatively charged overall (Cl-). Opposites attract and, voila, you have a chemical bond. And some salt – sodium chloride (NaCl).

By studying the Periodic Table we begin to see patterns in how easily electrons are won and lost, and realize that it's the distribution of all this negativity that determines how atoms stick together. The way in which electrons are won, lost or shared, determines the types of bonding that go on between atoms and the types of compounds these atoms make.

LIVING SITUATIONS

There are three main types of chemical bonds. Let's start with covalent bonding, where each molecule within a compound is a family of atoms that shares some electrons (see Single, double and triple bonds, right). These electrons are only shared among members of the same molecule. Think of it as a living situation – each molecule, or family, lives in a nice detached house, holding on to its own stuff and keeping itself to itself. This is

Single, double and triple bonds

Simply put, each covalent bond is a shared pair of electrons. The number of electrons that an atom has to share is usually the same as the number in its outer shell. So, for example, because carbon dioxide has four atoms to share, it can form up to four shared pairs, or four bonds. This idea of carbon forming four bonds is important in the structures of nearly all organic (carbon-containing) compounds, in which carbon skeletons are decorated with other types of atoms - in long-chain organic molecules, for example, carbon atoms share their electrons with each other and also, often, with hydrogen atoms. But sometimes, atoms share more than one pair with another atom. So you can have a carbon-carbon double bond or a carbon-oxygen double bond. You can even have triple bonds, where atoms share three pairs of electrons, though not all atoms have three electrons to share. Hydrogen, for instance, only has one.

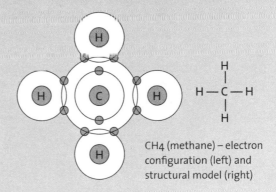

CH4 (methane) – electron configuration (left) and structural model (right)

1912

Concept of hydrogen bonding developed by Tom Moore and Thomas Winmill, later credited by Linus Pauling.

1939

The Nature of the Chemical Bond by Linus Pauling published

1954

Pauling awarded Nobel Prize for Chemistry for work on chemical bonding

2012

Quantum chemists propose a new chemical bond that occurs in very strong magnetic fields, such as in dwarf stars

how molecules like carbon dioxide, water and ammonia – the smelly compound used in fertilizers – live.

Ionic bonds, meanwhile, work by the 'opposites attract' model of bonding, as with sodium chloride in the previous example of salt. This type of bonding is more like living in a block of flats, where each occupant has neighbours either side, as well as above and below. There are no separate houses – it's just one big high-rise apartment block. The occupants, for the most part, keep a hold on their own stuff but close neighbours give and take the odd electron. This is what bonds them together – in ionically bonded compounds, the atoms stick together because they exist as oppositely charged ions (see Ions, page 19).

Then there's metallic bonding. Bonding in metals is slightly stranger. It works on the same principle of opposite charges being attracted to each other but, instead of a high-rise block, it's more like a hippie commune. All the electrons are shared communally. The negatively charged electrons float around, being picked up and shrugged off by the positively charged metal ions. Since everything belongs to everyone, there's no stealing – it's as if the whole thing is held together by trust.

These bonds aren't enough to hold the whole Universe together though. As well as the strong bonds within molecules and compounds, there are weaker forces that hold whole collections of molecules together – like the social ties that bind communities together. Some of the strongest of these forces are observed in water.

WHY WATER IS SPECIAL

You might never have considered it but the fact that the water in your kettle boils at 100 degrees centigrade is pretty odd. Water's boiling temperature is much higher than we'd expect for something composed of hydrogen and oxygen. We might reasonably assume from a study of the Periodic

Table (see pages 204–5) that oxygen would behave similarly to other elements occupying the same column. However, if you made hydrogen compounds with the three elements below oxygen, you certainly wouldn't be able to do something as simple as boil them in a kettle. That's because all three boil at temperatures below zero degrees (centigrade), which means they are gases at the temperature of your kitchen. Below zero, water is still solid ice. So why does a compound of oxygen and hydrogen stay liquid to such a high temperature?

Van der Waals

Van der Waals forces, named after a Dutch physicist, are very weak forces between all atoms. They exist because even in stable atoms and molecules, electrons shift around a little bit, changing the distribution of charge. This means that one negatively charged part of a molecule might temporarily attract a positively charged part of another. More permanent separations of charge occur in 'polar' molecules, such as water, allowing for slightly stronger attractions. Hydrogen bonding is a special case of this type of attraction, forming particularly strong intermolecular bonds.

The answer lies in the forces holding the water molecules together as a group, preventing them from flying off as soon as they feel a little heat. These so-called 'hydrogen bonds' form between the hydrogen atoms in one molecule and the oxygen atoms in another. How? Once again, it comes back to the electrons. In a water molecule, the two hydrogens find themselves in bed with sheet-stealing oxygen, which gathers up all the covers – negatively charged electrons – for itself. The now partially positive charges of the bare hydrogens mean they are attracted to sheet-stealing oxygens from other water molecules, which are more negative. As every water molecule has two hydrogens, it can form two of these hydrogen bonds with other water molecules. The same sticky forces help to explain the lattice structure of ice and the tension on the surface of a pond that allows a bug to skate across it.

The condensed idea
Sharing electrons

06 Changing phases

Few things stay the same. Chemists talk about transitions between different phases of matter, but that's really just a fancy way of saying that substances change. Matter can take multiple forms – as well as the everyday solid, liquid and gas states, there are some more unusual phases of matter.

Think about what happens if you leave a few squares of chocolate in your pocket on a hot day. You can take it out of your pocket and leave it in a cool place to harden again, but it won't taste quite the same as it originally did. Why? The answer lies in understanding the difference between the original chocolate and the re-hardened chocolate. First we're going to have to go back to school science lessons.

SOLIDS, LIQUIDS AND GASES ... AND PLASMA

There are three phases of matter that most people have heard of: solids, liquids and gases. Remember when you learned these in school? You probably remember them as 'states'. A basic example of a substance changing state is water freezing and melting – switching between a solid and a liquid state. Many other substances melt too, turning from solid to liquid, and chemists refer to this as thermal fusion. The different states are often explained by the packing of the atoms or molecules in the substance. In a solid, they are crammed together tightly like people in a crowded lift, whereas in a liquid the molecules circulate and move around each other more freely. In the gaseous state, the particles are more spread out and have

TIMELINE

1832	1835	1888
First use of melting points to characterize organic compounds	Adrien-Jean-Pierre Thilorier publishes first observation of dry ice	Liquid crystals discovered by Friedrich Reinitzer

no fixed boundaries – it's as if the lift doors have opened and the passengers have drifted off in all different directions.

These three states of matter mark the limits of many people's knowledge, but there are several more that are somewhat more esoteric and perhaps less well known. First off, there is the futuristic-sounding plasma. In this gaslike state – used in plasma TV screens, for example – the electrons have come loose and the particles of matter have become charged. What's different here, to continue the analogy, is that when the lift doors open, everyone moves off together, in a more orderly manner. Because the particles are charged plasma flows rather than bouncing around all over the place. Liquid crystals – used in LCD TVs – are yet another weird state of matter (see Liquid Crystals, page 26).

> **[IT] GLIDES RAPIDLY OVER A POLISHED SURFACE, AS IF IT WERE RAISED BY THE GASEOUS ATMOSPHERE WHICH CONSTANTLY SURROUNDS IT, UNTIL IT ENTIRELY DISAPPEARS.**
>
> French chemist Adrien-Jean-Pierre Thilorier on his first observations of dry ice

MORE THAN FOUR

Four states, or phases, might be adequate to understand many of the changes we see in substances in everyday situations. It can even explain some of the less everyday ones. For example, the smoke machines used in theatres and nightclubs that create very dense clouds of smoke or fog are using 'dry ice', which is frozen or solid carbon dioxide (CO_2). When it is dropped into hot water it does something quite unusual: it turns straight from a solid to a gas without passing through the liquid phase. (This, incidentally, is why it's called dry ice). The phase change from solid to gas is called sublimation. As soon as it happens, the still-cold bubbles of gas condense water vapour in the air, producing a fog.

Four phases, though, still doesn't explain the question posed at the top of the page – why the same chocolate can taste different, just because it has melted and then solidified. It's still a solid after all. But that's because there are more phases than the three or four classic states of matter can account for. Plenty of

1928	**1964**	**2013**
The term plasma is coined by Irving Langmuir	First working liquid crystal displays	New phase of water predicted to be present in 'ice-giant' planets

Liquid crystals

The liquid crystal state is one that most of us have already heard of because of its application in the liquid crystal displays (LCDs) that are used in modern electronic devices. Many different materials exhibit this state and not just the materials in your TV – the chromosomes in your cells can be thought of as liquid crystalline too. As its name suggests, the liquid crystal state is somewhere between a liquid and a solid crystal. The molecules, which are usually rod-shaped, are randomly arranged in one direction (as in a liquid) but regularly packed (as in a crystal) in another. This is because the forces holding the molecules together are weaker in one direction than in the other. The molecules in liquid crystals form layers that can slide over each other. Even within the layers, the randomly arranged molecules still move around. It's this combination of movement and regular arrangement that allows crystals to behave like liquids. In LCD screens, the positions of the molecules and the spaces between them affect how they reflect light and what colour we see. By using electricity to affect the positions of liquid-crystal molecules sandwiched between glass, we can create patterns and images on a screen.

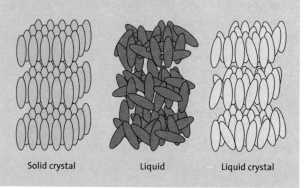

| Solid crystal | Liquid | Liquid crystal |

substances have multiple phases within the solid state and many of these solid forms are made up of crystals. The cocoa butter in chocolate, in fact, is crystalline and differences in how its crystals form determine the phase that it is in.

SIX TYPES OF CHOCOLATE

Finally then, we're ready to tackle the tasty topic of chocolate. You may have begun to wonder by now whether chocolate is perhaps a bit more complicated than it looks. It is. The main ingredient, cocoa butter, is made of molecules called triacylglycerols, but to keep this simple we'll just refer to them as 'cocoa butter'. The cocoa butter crystallizes in no less than six different forms or polymorphs, all of which have different structures and melt at different temperatures. Melting and then allowing your chocolate to re-harden gives you a different polymorph, each with a unique taste.

Even if you keep your chocolate at room temperature, it will slowly but surely morph into a different form – the most stable polymorph. Chemists call this change a phase transition, and this explains why you can sometimes unwrap a bar of chocolate that's been hanging around for a few months to discover that it looks as if it's diseased. The white bloom won't do you any harm. It's just polymorph VI. In a sense, all cocoa butter 'wants' to be polymorph VI because it's the most stable form. But it doesn't taste so great. To avoid the slow transition to VI, you can try keeping your chocolate at a lower temperature – in the fridge, for example.

Being able to manipulate the different forms of chocolate is obviously of great interest to the food industry and some very sophisticated studies on chocolate polymorphs have been carried out just in the last few years. In 1998, the chocolate manufacturer Cadbury even employed a particle accelerator to probe the secrets of tasty chocolate, using it to work out the different forms of crystal cocoa butter and how to make the best melt-in-the-mouth confection.

New phases

Substances can exist in multiple phases, and there are even phases that haven't been discovered yet. It seems scientists are constantly encountering new phases of water (see page 116). In 2013, a paper published in the scientific journal, *Physical Review Letters*, announced a new type of superstable, 'superionic' ice that is predicted to be present in large quantities in the cores of ice-giant planets like Uranus and Neptune.

The delicious, glossy-looking form that all of us want to eat is polymorph V, but getting a whole slab to crystallize in the V form is not easy. It requires a highly controlled process of melting and cooling at specific temperatures, in order to get the crystals to form in the correct way. Most importantly, of course, you have to eat it before it changes phase again. So now, boys and girls, you have a really good excuse for eating all your Easter eggs by Easter Monday!

The condensed idea
Not just solids,
liquids and gases

07 Energy

Energy is like some sort of supernatural being: powerful but unknowable. Although we can witness its effects, it never reveals its true form. In the 19th century, James Joule laid the foundations for one of the most fundamental laws in science. This law governs the energy changes that occur in every chemical reaction.

If you were playing a game of charades and had to come up with a mime for energy, what would it be? It's a puzzle because energy is pretty difficult to pin down. It's fuel, it's food, it's heat, it's what you get out of solar panels; it's a coiled spring, a falling leaf, a billowing sail, a magnet, a thunderbolt and the sound of a Spanish guitar. If energy can be all of these things, then what is its essence?

WHAT IS ENERGY?

All living things use energy to build their bodies and grow, and, in some cases, to move about. Humans seem addicted to the stuff – harnessing vast quantities of it to light our homes, and fuel our technology and power our factories. However, energy is not a substance as we recognize it – we can't see it or get our hands on it. It's intangible. We've always been aware of its effects, if only vaguely, but it's only since the 19th century that we've really known it exists. Prior to the work of English physicist James Prescott Joule, we had only a fuzzy idea of what energy actually was.

TIMELINE

1807	1840	1845
Term 'energy' coined by Thomas Young	Joule's Law relates heat to electric current	Joule first reports on paddle wheel experiments in presentation of *On the Mechanical Equivalent of Heat*

Joule was a brewer's son who was home-schooled and conducted many of his experiments in the cellar of the family brewery. He was concerned with the relationship between heat and movement – so concerned was he that he took his thermometers (and William Thomson) with him on his honeymoon in order to study temperature differences between the top and bottom of a nearby waterfall! Joule had trouble getting his early papers published, but thanks to some famous friends – not least electrical pioneer Michael Faraday – he did eventually succeed in getting his work noticed. His key insight was essentially this: heat is movement.

Work

Although energy is very tricky to define, it can be thought of as the ability to produce heat or 'do work'. Admittedly, this sounds somewhat ambiguous. Do work? What work? Work is actually an important concept in physics and chemistry related to movement. If anything moves, then work is being done. A combustion reaction, like in a car engine, releases heat, which moves the pistons (doing the 'work') as gases in the engine expand.

Heat is movement? At first reading, this observation might not seem to make much sense. But think about it. Why do you rub your hands together to warm them up on a cold day? Why do the tyres of a moving vehicle get hot? Joule's paper *On the Mechanical Equivalent of Heat*, published on New Year's Day 1850, asked the same sort of questions. In it, he noted that the sea becomes warm after days of stormy weather and detailed his own attempts to replicate the effect using a paddle wheel. By taking accurate temperature measurements using his trusty thermometers, he showed that motion can be converted to heat.

Through Joule's research and work by German scientists Rudolf Clausius and Julius Robert von Mayer, we learned that mechanical force, heat and electricity are all very closely related. The Joule (J) eventually became a standard unit by which to measure 'work' (see Work, above) – a physical quantity that can be thought of as energy.

1850
Expanded version of *On the Mechanical Equivalent of Heat* published in *Philosophical Transactions of the Royal Society of London*

1850
Rudolf Clausius and William Thomson state the First and Second Law of Thermodynamics

1905
Albert Einstein's $E = mc^2$ relates energy (E) to mass (m) and speed of light (c)

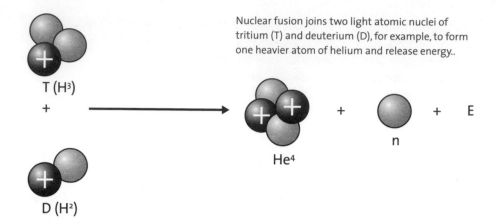

T (H³)

+

D (H²)

Nuclear fusion joins two light atomic nuclei of tritium (T) and deuterium (D), for example, to form one heavier atom of helium and release energy..

He⁴   +   n   +   E

FROM ONE TO ANOTHER

Today we recognize many different types of energy, and understand that one can be converted to another. The chemical energy in coal or oil, for instance, is stored energy until it is burned and turned into heat energy to warm our houses. So the connection that Joule made between heat and movement no longer seems so odd, since we now consider both as types of energy. At a deeper level, though, heat really *is* movement – what makes a pan of hot water hot is the fact that the energetic water molecules in it are all jiggling around in an excited state. Movement is just another type of energy.

In chemicals, energy is stored in the bonds between atoms. When bonds are broken in chemical reactions energy is released. The opposite process, the forming of bonds, puts the energy in storage for later. Like the energy in a coiled spring, that energy is 'potential energy', available until it is released. Potential energy is just energy that is stored in an object due to its position. With chemical potential energy, this refers to the position of bonds. When you stand at the top of the stairs, your potential energy is greater than when you stand at the bottom, and there was also potential energy in the water at the top of Joule's honeymoon waterfall. Your potential energy depends on your mass – were you to sit and eat cakes for a month and then go and stand at the top of the stairs once again, your potential energy would be increased.

Even sitting and eating cakes is an example of an energy change – the sugar and fat in the cakes provide chemical energy that is converted by your cells to heat energy for maintaining your body temperature and movement energy for driving the muscles that will take you to the top of the stairs. Everything we do, everything our bodies do and basically everything that ever happens relies on these energy conversions.

ENERGY CHANGES BUT STAYS THE SAME

James Joule's work laid the foundations for what has become one of the most important principles in all of science: the Law of the Conservation of Energy, otherwise known as the First Law of Thermodynamics (see page 40). This law states that energy is never created and never destroyed. It is only converted from one form to another – as hinted by Joule's paddle-wheel experiments. Whatever happens in chemical reactions, and anywhere else, the total amount of energy in the Universe must alway remains the same.

> **MY OBJECT HAS BEEN, FIRST TO DISCOVER CORRECT PRINCIPLES AND THEN TO SUGGEST THEIR PRACTICAL DEVELOPMENT.**
>
> James Prescott Joule,
> *James Joule: A Biography*

What all energy has in common is its ability to change something. Now, whether this tells you how to mime energy in a game of charades is quite another matter. Energy is a paddle wheel rotating. It's a cake. It's you walking up the stairs, you standing at the top of the stairs and you falling down the stairs. Try miming those. It's just as confusing as it ever was.

The condensed idea
The ability to make a change

08 Chemical reactions

Chemical reactions aren't just the noisy explosions that might fill the air of a cartoon scientist's lab. They're also the everyday processes that quietly tick along inside the cells of living things – including us. They happen without us even noticing. Still, we all love a good noisy explosion!

There are, very broadly speaking, two types of chemical reactions. There is the big, flashy, explosive type of chemical reaction – the 'stand-well-back-and-wear-goggles' type – and then there is the quiet, 'plod-along, barely-notice-it' type of reaction. The stand-well-back type might grab our attention but the barely-notice-it type can be just as impressive. (In reality, of course, there is a dizzying variety of different chemical reactions and far too many to list here.)

Chemists are suckers for the first type. But aren't we all? Who, given a free ticket to a fireworks display, would rather sit quietly and watch rust forming? Who didn't jump and giggle just a little bit when their chemistry teacher set fire to a hydrogen balloon, producing a loud BOOM? If you ask any chemist to demonstrate their favourite reaction, they will invariably conjure up the biggest and flashiest experiment that they can safely pull off. To begin to understand chemical reactions, we turn to a nineteenth-century chemistry teacher, and one of the noisiest and most spectacular chemistry demonstrations. Unfortunately, these types of experiment do not always go to plan.

TIMELINE

1615	1789	1803
First equation-like reaction scheme	Concept of chemical reactions emerges from Antoine-Laurent Lavoisier's *Elementary Treatise on Chemistry*	John Dalton's atomic theory proposes chemical reactions as rearrangements of atoms

STAND WELL BACK

Justus von Liebig was an extraordinary person. He lived through a famine, became a professor aged 21, uncovered the chemical basis of plant growth and founded a leading scientific journal, not to mention making some discoveries that led to the invention of yeast extract spread (aka Marmite). He did many things that he could be proud of, but he also did some things to be embarrassed about. Legend has it that while demonstrating a reaction known as the 'Barking Dog' for the Bavarian royal family in 1853, his experimented exploded a little too violently – right in the face of Queen consort Therese of Saxe-Hildburghausen and her son, Prince Luitpold.

The Barking Dog is still one of the most spectacular scientific demonstrations. It's not just fantastically explosive and noisy – emitting a loud 'woof' – it's also brilliantly flashy. The reaction happens when carbon disulfide (CS_2) is mixed with nitrous oxide (N_2O) – better known as laughing gas – and set alight. It's an exothermic reaction, meaning it loses energy to its surroundings (see page 30). In this case, some of the energy is lost in the form of a big, blue flash of light. Performed, as it often is, in a large, transparent tube, the experiment is reminiscent of a light sabre being 'ignited' and then switched off. It's worth taking a moment to track down an online video if you can.

> **...I LOOKED AROUND AFTER THE TERRIBLE EXPLOSION IN THE ROOM ... AND SAW THE BLOOD RUNNING FROM THE FACES OF QUEEN THERESE AND PRINCE LUITPOLD.**
>
> Justus von Liebig

Had Liebig's audience not been so impressed by the effect, they wouldn't have convinced him to repeat it, and Queen Therese wouldn't have ended up sufffering her minor injury – the explosion is said to have drawn blood. Like all reactions though, the Barking Dog is just a rearrangement of atoms. There are only four different types of atoms – elements – involved in the Barking Dog: carbon (C), sulfur (S), nitrogen (N) and oxygen (O).

1853	1898	1908	2013
Bavarian queen injured by famous 'Barking Dog' reaction	Term photo-synthesis used to describe photo-synthetic reactions	Fritz Haber sets up a pilot plant for producing ammonia from nitrogen and hydrogen	Atomic force microscopy used to watch reactions happening in real time

Barking Dog reaction: in a similar, parallel reaction, CO2 may also be formed

$$N_2O + CS_2 \longrightarrow N_2 + CO + SO_2 + S_8$$

Chemists use a chemical equation to show where they end up after the reaction.

BARELY NOTICE IT

But what of the quieter, less showy reactions? The gradual rusting of an iron nail is a chemical reaction between iron, water and oxygen in the air, to form the iron-oxide product – orange-brown, rust flakes (see page 52). It's a slow oxidation reaction. When you cut an apple and it turns brown, that's another oxidation reaction – one that you can watch over the course of a few minutes. For one of the most important quiet reactions, look no further than the plants in your window box. They silently harvest the Sun's rays and use the energy to rearrange carbon dioxide and water into sugar and oxygen, in a reaction we know as photosynthesis (see page 148). This is a summary of a much more complex chain of reactions evolved by plants. The sugar is used as fuel to power life in the plant, while the other product, oxygen, is released. It might not be as dramatic as in the Barking Dog, but it's central to life on our planet.

Chemical equations

In 1615, Jean Beguin published a set of chemistry lecture notes showing a diagram of the reaction of *mercure sublimé* (mercury chloride, $HgCl_2$) with *antimoine* (antimony trisulfide, Sb_2S_3). Although it looks more like a spider diagram, it is considered an early representation of a chemical equation. Later, in the eighteenth century, William Cullen and Joseph Black, who lectured at Glasgow and Edinburgh universities, drew reaction schemes containing arrows to explain chemical reactions to their students.

You can look to your own body for examples of reactions. Your cells are essentially bags full of chemicals, miniature reaction centres. Each one does the opposite of what a plant does in photosynthesis – to release energy, it reacts the sugar absorbed from your food with the oxygen that you breathe in and rearranges them, producing carbon dioxide and water. This mirror-image 'respiration reaction' is the other big life-sustaining reaction on Earth.

Watching reactions unfold

Usually, when we say we 'see' a reaction taking place, we are just referring to the explosion, the colour change or some other consequence of the reaction. We're not looking at the individual molecules, so we can't actually see what's going on. But in 2013, US and Spanish researchers really did see reactions happening in real time. They harnessed the power of atomic force microscopy to take extremely close-up images of single oligo–(phenylene-1, 2-ethynylene) molecules reacting on a silver surface to form new ring-structured products. In atomic-force microscopy, the images are generated in a completely different way to those in a normal camera. The microscope has a very fine probe or 'tip', which produces a signal when it touches something on a surface. It can detect the presence of single atoms. In the images taken in 2013, the bonds as well as the atoms in the reactants and products are clearly visible.

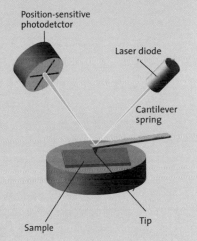

Position-sensitive photodetctor

Laser diode

Cantilever spring

Sample

Tip

REARRANGEMENTS

Whether they are big or small, slow or gone in a flash, all reactions are the result of some change in the way that the atoms in the starting reactants are arranged. The atoms of the various elements may be ripped apart and stuck back together in different ways. This usually means that new compounds are formed – held together by the sharing of electrons between new partner atoms. In the Barking Dog reaction, carbon monoxide and sulfur dioxide are the two new compounds formed. Nitrogen and sulfur molecules are also produced. In photosynthesis, larger, more complex molecules – long sugar molecules containing multiple carbon, hydrogen and oxygen atoms – are formed.

The condensed idea
Rearranging atoms

09 Equilibrium

Some reactions head in only one direction, while others flip backwards and forwards constantly. In these 'flexible' reactions, an equilibrium keeps the status quo. Equilibrium reactions are found everywhere, from your blood to the fuel system that brought the Apollo 11 astronauts back to Earth.

You have some friends coming over and have bought a couple of bottles of red wine. Keen to get the party started, you crack open a bottle, pour out a few glasses and wait for everyone to arrive. An hour later, following a barrage of text-message apologies, you and your one friend are still sipping your first glasses of wine, while all the others stand untouched. One of two things might happen now. Either your friend will make some polite excuse, leaving you to pour the untouched glasses of wine back into the bottle. Or the pair of you finish both glasses, plus the others you poured, and then open the next bottle and start pouring some more.

KEEPING THE WINE FLOWING

You might be wondering what all this has to do with chemistry. Well, there are many reactions in chemistry that mirror the wine situation at the failed party. Just like the action of pouring wine from a bottle to glass and back into the bottle again, these reactions are reversible. In chemistry, this type of situation is called equilibrium, and the equilibrium controls the proportions of reactants and products of a chemical reaction.

Imagine that the bottled wine represents the chemical reactants, while the wine poured into glasses represents the products of the reaction. At your

TIMELINE

1000	1884	1947
The Great Stalactite started to form	Le Châtelier's Principle	Paul Samuelson applies Le Châtelier's Principle to economics

party, you control the flow of the wine, so if someone drinks a glass you pour out another one. In the same way, the equilibrium controls the flow from reactants to products, so if some of the products disappear, it works to find the status quo by turning some of the reactants into new products. But a reversible reaction also works the opposite way, so if something interferes with the status quo and suddenly there are too many products, the equilibrium just pushes the reaction back in the other direction and converts the products back into reactants – just like pouring wine back into a bottle.

The existence of an equilibrium doesn't mean that each side of the chemical equation is equal – there isn't always the same volume of wine in the glasses as there is in the bottle. Instead, each chemical system has its own happy medium, where the forward and backward reactions happen at the same rate. This applies not just to complex reactions, but to simple systems, such as weak acids (see page 45), donating and accepting hydrogen (H^+) ions, and even water molecules

Equilibrium constant

Each chemical reaction has its own equilibrium, but how do we know where it is? Something called the equilibrium constant determines what proportion of the reactants is transformed into products in a reversible reaction – it tells us where the equilibrium lies. The equilibrium constant has the symbol K and its value is the same as the ratio of products to reactants. So if there are equal amounts (or concentrations) of products and reactants, then K is equal to 1. However, if there are more products then K is higher than 1, and if there are more reactants then K is lower than 1. Every reaction has its own value for K. In the production of industrial chemicals, catalysts are used to modify the equilibrium constant, pushing it to create more products. Reactions that are carried out to make useful chemicals, such as ammonia (see page 68), must constantly readjust themselves to balance out the removal of the products. This is because removing the products temporarily changes the ratio of products to reactants, or K. To maintain K, the reaction must drive slightly harder in the forward direction, making more products again.

$$A \rightleftharpoons B$$
$$\text{reactants} \rightleftharpoons \text{products}$$

$$K_{eq} = [B] / [A]$$

(square brackets = concentration)

1952

The Great Stalactite discovered

1969

Nitrogen tetroxide blasts the Apollo 11 crew back to Earth

splitting up into H^+ and OH^- ions. In water, the equilibrium lies closer to the H_2O of the system, than to the separate ions, so whatever happens the equilibrium will work to keep most of the water as H_2O molecules.

ROCKET FUEL

So where else do we find these sorts of chemical equilibria? The 1969 Moon landing provides a good example. Designed by NASA, the system that allowed Neil Armstrong, Buzz Aldrin and Michael Collins to return home from the Moon was a chemical one. To generate the thrust that blasted them back into space, they needed a fuel and an oxidizing agent – something that makes the fuel burn more fiercely by adding oxygen to the mix. The oxidizing agent used in the Apollo 11 mission was called dinitrogen tetroxide (N_2O_4) a molecule that splits down the middle to form two nitrogen dioxide molecules (NO_2). But the NO_2 can convert easily back into N_2O_4. Chemists show it like this:

> **THERE EXISTS EVERYWHERE A MEDIUM IN THINGS, DETERMINED BY EQUILIBRIUM.**
> Dmitri Mendeleev

$$N_2O_4 \rightleftharpoons 2NO_2$$

If you put nitrogen tetroxide in a glass jar (not advisable as it's corrosive and if you spill any you'll lose some skin), you'll see the equilibrium at work. When it's kept cold, the brownish dinitrogen tetroxide sits at the bottom of the jar, while the NO_2 molecules hang in a vapour cloud above. However, temperature and other conditions can change the balance of an equilibrium. In the case of dinitrogen tetroxide, a little heat drives the equilibrium towards the right-hand side of the equation, turning more of the oxidizing agent into gas. Cooling it again converts some of it back into N_2O_4.

NATURAL BALANCE

Equilibria occur in nature all the time. They keep chemicals in your blood in check, maintaining a constant pH around 7, ensuring your blood never becomes too acidic. Linked to these same equilibria are reversible reactions that control the release of carbon dioxide in your lungs. You then breathe out the carbon dioxide.

If you've ever seen the drips and spires of stalactites and stalagmites that form in limestone caves, you might have pondered how they form. The Great Stalactite that hangs from the ceiling of the Doolin Cave on the west coast of Ireland is one of the largest in the world at over seven metres long. It formed over thousands of years. This natural wonder is in fact another example of a chemical equilibrium in action.

$$CaCO_3 + H_2O + CO_2 \rightleftharpoons Ca_2^+ + 2HCO_3^-$$

$CaCO_3$ is the chemical formula for calcium carbonate, which forms the porous rock, limestone. Rainwater that has carbon dioxide dissolved in it produces a weak acid called carbonic acid (H_2CO_3), which reacts with the calcium carbonate in limestone, dissolving it to produce calcium and hydrogen carbonate ions. As the rain penetrates through the holes in the rock, it dissolves bits of the limestone and carries the dissolved ions with it. This slow process is enough to create vast limestone caverns. Stalactites, such as the Great Stalactite, form where this water containing calcium and hydrogen carbonate ions drips in the same place for a long time. As the rainwater drips, the opposite reaction occurs. The ions are converted back into calcium carbonate, water and carbon dioxide, and limestone is deposited. Eventually, the continuous build-up of limestone at the drip creates a drip of solid rock in its image, with stunning results.

Le Châtelier's Principle

In 1884, Henri Louis Le Châtelier proposed a governing principle of chemical equilibria: 'Every system in chemical equilibrium, under influence of a change of every single one of the factors of equilibrium, undergoes a transformation in such direction that, if this transformation took place alone, it would produce a change in the opposite direction of the factor in question.' In other words, when a change occurs in one of the factors influencing an equilibrium, the equilibrium adjusts to minimize the effects of the change.

The condensed idea
Status quo

10 Thermodynamics

Thermodynamics is a way of predicting the future for chemists. Based on some fundamental laws, they can work out whether something is going to react, or not. If it is hard to get excited about thermodynamics, consider that it has much to say concerning tea and the end of the Universe.

Thermodynamics might sound like one of those crusty old subjects that no one really needs to know about these days. It is, after all, based on scientific laws developed over a hundred years ago. What can thermodynamics possibly teach us today? Well, quite a lot, actually. Chemists are using thermodynamics to figure out what happens in living cells when they get cold – for example, when human organs are packed in ice before they are transplanted. Thermodynamics is also helping chemists to predict the behaviour of liquid salts that are being used as solvents in fuel cells, drugs and cutting-edge materials.

The Laws of Thermodynamics are so fundamental to the business of science that we'll always be finding new ways to work with them. Without the Laws of Thermodynamics, it would be hard to understand or predict why any chemical process or reaction happens in the way that it does. Or to rule out the possibility that ordinary processes might happen in some other crazy way – like your cup of tea getting hotter the longer you leave it to stand. So what are these indisputable laws?

TIMELINE

1842	1843	1847
Julius Robert Mayer formulates law of conservation of energy	James Prescott Joule also formulates law of conservation of energy	Hermann Ludwig von Helmholtz formulates law of conservation of energy again

ONE NEITHER CREATES NOR DESTROYS

We've already met the First Law of Thermodynamics (see page 31). In its simplest form, it states that energy can never be created or destroyed. This only makes sense if we remember what we know about energy conversions: that energy can be converted from one form to another, for example, when the chemical energy in the fuel tank of your car is converted into kinetic, or movement, energy after you switch the engine on. It's these energy conversions that people who study thermodynamics tend to be really interested in.

Chemists might say that energy is 'lost' during a particular chemical reaction, but it is not really lost. It has just gone somewhere else – to the surroundings, as heat, usually. In thermodynamics, this type of 'heat-loss' reaction is referred to as exothermic. The opposite, a reaction that absorbs heat from its surroundings, is called endothermic.

The important thing to remember is that no matter how much energy is transferred between the materials that are part of the reaction and their surroundings, the total energy always stays the same. Otherwise,

Systems and Surroundings

Chemists like to be ordered about things so when they are doing their thermodynamic calculations they always make sure that they have categorized what they're talking about. The first task is always to identify the specific system or reaction that they're studying, and then everything else is the surroundings. A cup of tea going cold, for example, must be thought of as the tea itself and then everything that surrounding the tea – the cup, the coaster, the air that the steam evaporates into, the hand that you warm on the hot mug. Actually, when it comes to chemical reactions, it can be more difficult than you might think to work out where the system ends and the surroundings begin.

A complete thermodynamic system

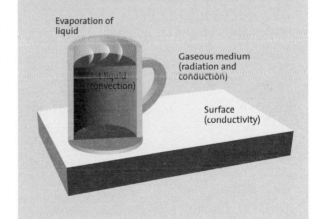

Evaporation of liquid

Gaseous medium (radiation and conduction)

Hot liquid (convection)

Surface (conductivity)

1850	1877	1912	1949	1964
Rudolf Clausius and William Thomson state the First and Second Laws of Thermodynamics	Ludwig Boltzmann describes entropy as a measure of disorder	Third Law of Thermodynamics stated by Walther Nernst	William Francis Giauque wins Nobel Prize for advances in chemical thermodynamics	Flanders and Swann release their song, *First and Second Law*

the principle of energy conservation – the First Law of Thermodynamics – wouldn't work.

THE SECOND LAW DESTROYS THE ENTIRE UNIVERSE

The Second Law of Thermodynamics is a little trickier to grasp, but it manages to explain pretty much everything. It has been used to explain the Big Bang and predict the end of the Universe, and – along with the First Law – tells us why attempts to build a perpetual-motion machine are doomed to failure. It also helps us to understand why tea goes cold instead of getting hotter.

The tricky part about this Second Law is that it relies on a difficult concept called entropy. Entropy is often described as a measure of disorder – the more disordered something is, the higher its entropy. Think of it like a packet of pretzels. When all the pretzels are safely in the packet, their entropy is pretty low. As you open the packet, far too eagerly, the pretzels explode all over the place, their entropy becomes much higher. The same is true if you uncork a vial of smelly methane gas – in this case, your nose will be able to detect the disorder coming.

> **NOT KNOWING THE SECOND LAW OF THERMODYNAMICS IS LIKE NEVER HAVING READ A WORK BY SHAKESPEARE.**
> C.P. Snow

The Second Law of Thermodynamics states that entropy always increases, or at least never decreases. In other words, things tend to get more disordered. This applies to everything including the Universe itself, which will eventually fall into complete disarray and expire. The rationale for this altogether chilling prediction is that, in essence, there are far more ways of throwing the pretzels around than there are of them staying in the bag (see Entropy, page 43). The Second Law is sometimes described in terms of heat, by saying that heat always flows from hotter places to colder places – thus your tea always loses heat to its surroundings and goes cold.

From a chemist's point of view, however, the Second Law is important for determining what happens in chemical processes and reactions. A reaction is only thermodynamically feasible, or in other words can only 'go' in a certain direction, if entropy increases overall. To work this out, the chemist must think not just about the entropy change in the 'system', which often turns out to be much more complicated than a bag of pretzels or a cup of tea, but also about the entropy change in its surroundings (see Systems and Surroundings, page 41). As long as the Second Law is not violated, the reaction can proceed and, should it not work, then the chemist has to figure out what needs to be done to make it work.

Entropy

What entropy actually measures is how many different states a system could exist in, given some key parameters. We might know the size of a bag of pretzels and even how many pretzels are in it; however, if we shake it up and down we don't know exactly where each pretzel will be when we open it. Entropy tells us how many different ways there are of arranging the pretzels. The bigger the bag, the more ways there are to arrange the pretzels. In chemical reactions, with molecules rather than pretzels, there are even more parameters to consider, such as temperature and pressure.

WHO'S AFRAID OF THE THIRD LAW?

The Third Law of Thermodynamics is less well known than the other two laws. What it essentially says is that when the temperature of a perfect crystal – and it must be perfect – hits absolute zero, its entropy should also be zero. And that perhaps explains why the Third Law of Thermodynamics is often forgotten about. It all seems a bit abstract and it is assumed to be useful only for people who have the ability to cool stuff to absolute zero (–273 °C) and who are working with crystals – and perfect and ideal crystals at that!

The condensed idea
Energy change

11 Acids

How come you can keep vinegar in a glass bottle, shake it on your chips and eat it, while fluoroantimonic acid would eat the bottle itself? It's all down to one tiny little atom that's found in every acid from the hydrochloric acid in your stomach to the world's strongest superacids.

Humphry Davy was a lowly surgeon's apprentice who became famous for encouraging well-to-do people to inhale laughing gas. Born in Penzance, Cornwall, and a literary man at heart, Davy became friends with some of the west of England's best-renowned romantic poets – Robert Southey and Samuel Taylor Coleridge – but it was in chemistry that he made his career. He accepted a job in Bristol as a chemical superintendent, where he published the work that would secure him a lectureship and eventually a post as Professor of Chemistry at the Royal Institution in London.

Nineteenth-century cartoons show Davy entertaining audiences at his lectures with bellows full of nitrous oxide – laughing gas – although he did propose that the therapeutic gas be used as an anaesthetic. Outside of his popular lectures, Davy was carrying out pioneering work in electrochemistry (see page 92). Although he was not the first to realize that electricity could split compounds into their component atoms, he put the technique to good use in discovering the elements potassium and sodium. He also tested a theory put forward by one of the big names in chemistry, Antoine Lavoisier.

Lavoisier had met his death – by guillotine – a few years before, at the hands of the French Revolution. Although he is remembered for many enlightening

TIMELINE

1778	1810	1838
Antoine-Laurent Lavoisier's oxygen theory of acids	Humphry Davy disproves the oxygen theory	Justus von Liebig's hydrogen theory of acids

insights, such as his suggestion that water is composed of oxygen and hydrogen, he did get at least one thing wrong. He proposed that oxygen, the element that he himself had named, was what made acids acidic. But Davy knew otherwise. Using electrolysis, he split muriatic acid into its elements and found that it contained only hydrogen and chlorine. The acid contained no oxygen. Muriatic acid is the acid that you will find sitting on the shelf in any chemistry lab and the acid in your stomach that helps digest your food: hydrochloric acid.

HYDROGEN NOT OXYGEN

In 1810, Davy concluded that oxygen could not be what defined an acid. It took nearly another century for the first, truly modern theory of acids to emerge, courtesy of the Swedish chemist, Svante Arrhenius. An eventual Nobel Prize winner. Arrhenius proposed instead that acids were substances that dissolved in water to release hydrogen, as positively charged hydrogen ions (H^+). He also said that alkaline substances (see Bases, page 46) dissolved in water to release hydroxide ions (OH^-). Although Arrhenius' definition of bases was later revised, his central premise – that acids are hydrogen donators – forms the basis of our understanding of acids.

WEAK AND STRONG ACIDS

Today, we think of acids as proton donators and bases as proton acceptors. (Remember that in this context a proton means a hydrogen atom that has been stripped of its electron to form an ion, so this theory simply states that acids donate hydrogen ions and bases accept them). The strength of an acid is a measure of how good the molecule is at donating its proton. Vinegar, or ethanoic acid (CH_3COOH), you shake on your chips is pretty weak, because at

Moles

Chemists have a curious concept of quantity. Often, instead of just weighing stuff, they want to know exactly how many particles of it there are. They call a certain number of particles – equal to the number of particles in 12 g of ordinary carbon – a 'mole'. So a bottle of acid labelled 1 M (1 molar) tells you it contains 6.02×10^{23} of acid in each litre. Thankfully you don't have to count out each particle. Substances are given a 'molar mass' - the weight that equals one mole.

> **I SHALL ATTACK CHEMISTRY, LIKE A SHARK...**
>
> Poet Samuel Taylor Coleridge, friend of Humphry Davy

1903	1923	1923
Svante Arrhenius wins Nobel Prize for work on acid chemistry	Johannes Broensted and Thomas Lowry independently propose hydrogen-based acid theories	Gilbert Lewis definition of acids

Bases

On the pH scale, a base is considered to be a substance with a pH above 7 – the midpoint of the scale, which usually spans from 0 to 14 (even though negative pHs and those above 14 do exist). A base dissolved in water is called an alkali. Alkaline substances include ammonia and baking soda. A 2009 study by Swedish researchers found that alkaline substances, as well as acidic ones like fruit juices, can damage your teeth. Which makes the old-fashioned logic of brushing with baking soda – to neutralize acids – seem a bit outdated. Because the pH scale works in a logarithmic fashion, each single-point increase means that a substance is ten times more basic, and vice versa. So a pH-14 base is ten times more basic than a pH-13 base and a pH-1 acid is ten times more acidic than a pH-2 acid.

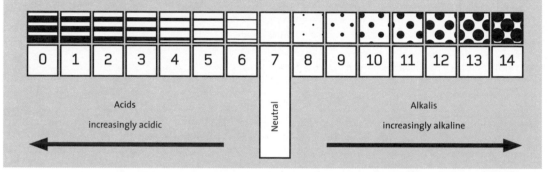

any one time a lot of the molecules will have their protons still attached. The protons are constantly splitting off and then joining back up with the main molecule again, forming an equilibrium mixture (see page 36).

Davy's hydrochloric acid (HCl), on the other hand, is really quite good at donating its protons. All of the hydrochloric acid dissolved in water is split into hydrogen and chlorine (Cl^-) ions – in other words, it ionizes completely.

The strength of an acid is separate from its concentration. Given the exact same number of acid molecules dissolved in the same amount of water, a stronger acid like hydrochloric acid will release more of its protons than a weaker acid and will therefore be at a higher concentration. However, one could dilute hydrochloric acid with enough water to make it less acidic than vinegar. Chemists measure the concentration of acids using the pH scale (see Bases, opposite). Confusingly, a lower pH means a higher concentration of hydrogen ions – a more concentrated acid is considered more acidic and has a lower pH number.

SUPERACIDS

The exciting thing about acids, as everyone knows, is that you can use them to dissolve all sorts of things – like desks, vegetables and, as popularized by the cult television show *Breaking Bad*, a whole dead body in a bath. Actually, the truth is that hydrofluoric acid (HF) would not burn straight through a bathroom floor or instantaneously reduce a body to mulch like it did in the TV show, although it certainly would hurt if you poured it on your hand.

If you want a really nasty acid, you can make it by taking hydrofluoric acid and reacting it with something called antimony pentafluoride. The resulting fluoroantimonic acid is so acidic that it falls off the bottom end of the pH scale. It is so violently corrosive that it has to be stored in Teflon – a material that is as tough as nails because it contains some of the strongest bonds (carbon–fluorine bonds) in all of chemistry. This acid is called a 'superacid'.

Some superacids will eat through glass. Oddly, though, carborane superacids, which are some of the most potent known, can be kept quite safely in an ordinary glass bottle. This is because it's not the bit that Arrhenius identified as being acidic – the hydrogen ion – that determines whether an acid is corrosive. It's the other component. It's the leftover fluorine in hydrofluoric acid that will corrode glass. In carborane superacids, which are stronger acids, the leftover part is stable and doesn't react.

The condensed idea
Setting hydrogen free

12 Catalysts

Some reactions just can't get along without help. They need a little push. Certain elements and compounds can act as helpers to provide that push and are called catalysts. In industry, catalysts are often metals and are used to drive reactions. Our bodies also use tiny amounts of metals, contained in molecules called enzymes, to speed up biological processes.

I n February 2011, doctors at the Prince Charles Hospital in Brisbane saw a 73-year-old woman with arthritis who complained of memory loss, dizziness, vomiting, headaches, depression and anorexia. None of her symptoms seemed to be related to the arthritis, or to the hip replacement she'd had five years earlier. After running some tests, the doctors realized the woman's cobalt levels were high. It turned out the metal joint of her new hip was leaking cobalt, resulting in her neurological symptoms. Cobalt is a toxic metal. It causes a rash on contact with skin and breathing problems when inhaled. In high doses, it can get you into all sorts of trouble. But we actually need cobalt to live. Like other transition metals (see page 8), such as copper and zinc, it is essential to the action of enzymes in the body. Its most crucial role is in vitamin B12, the vitamin that's found in meat and fish, and used to fortify cereals. It works, essentially, as a catalyst.

HELPING OUT

What is a 'catalyst'? You've probably heard it in relation to catalytic converters in cars (see Photocatalysis, page 51) or phrases like 'catalyst for innovation'.

TIMELINE

1912

Paul Sabatier receives
Nobel Prize for Chemistry
for work on metal catalysis

1964

Dorothy Hodgkin receives
Nobel Prize for Chemistry
for first metalloenzyme
structure

You've a vague notion that it means getting something started. But to understand what a chemical catalyst or a biological enzyme (see page 132) actually does, think of it as a helper particle. If you really need to paint the ceiling but it just seems like far too much effort, you might trespass on the sweet nature and expert DIY skills of a loved one or housemate to kickstart the process. You send them off to buy the right type of paint and a roller while you try to muster the energy to make it happen. It seems a bit easier now someone is giving you a hand.

Catalytic converters

The catalytic converter in a car is the part that removes the most harmful pollutants from the car's exhaust emissions – or at least converts them into other less harmful pollutants. Rhodium, a metal that's rarer than gold, has its main application in catalytic converters. It helps convert nitrogen oxides into nitrogen and water. Palladium is often the catalyst used to help convert carbon monoxide into carbon dioxide. So we may get carbon dioxide emissions, but at least we don't get carbon monoxide, which is a whole lot more deadly to people. In a catalytic converter, the reactants are gases, so the rhodium catalyst is said to be in a different phase (see page 32) to the reactants. These types of catalysts are called heterogeneous catalysts. When a catalyst is in the same phase as the reactants, it's referred to as a homogeneous catalyst.

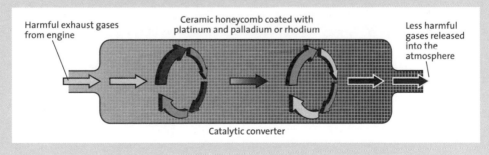

Harmful exhaust gases from engine

Ceramic honeycomb coated with platinum and palladium or rhodium

Less harmful gases released into the atmosphere

Catalytic converter

1975

First catalytic converters installed in cars

1990

Richard Schrock makes efficient metal catalyst for metathesis reactions

2001

Pilkington launches first self-cleaning glass based on photocatalysis

The same sort of thing happens in some chemical reactions. They just can't get going without some extra help but, a bit like when your housemate gives you a hand with the painting, the catalyst makes it all seem like a lot less effort. Actually, a catalyst really does reduce the amount of energy needed to get a reaction started – it creates a new route for the reaction that doesn't present such a big energy barrier for the reactants to overcome. As a bonus, it's not consumed by the reaction so it can help out again and again.

JUST A LITTLE BIT

> **NICKEL APPEARED … TO POSSESS A REMARKABLE CAPACITY TO HYDROGENATE ETHYLENE WITHOUT … BEING VISIBLY MODIFIED, I.E. BY ACTING AS A CATALYST.**
> Paul Sabatier, Nobel Prize for Chemistry, 1912

In the body, transition metals are often used by vitamins for their catalytic properties. B12 was for a long time the mystery factor gained by eating liver – 'the liver factor' – that could cure anaemic dogs and people. Helped by cobalt, it catalyses a number of different reactions that are important in metabolism and the manufacture of red blood cells. Its complex structure was the first among metalloenzymes to be discovered by X-ray crystallography (see page 88) in a series of painstaking analyses that won Dorothy Crowfoot Hodgkin the Nobel Prize for Chemistry in 1964. Other enzymes that carry onboard transition metal helpers include cytochrome oxidase, which uses copper to extract the energy from food in plants and animals.

Only the tiniest amount of cobalt is needed to keep the few milligrams of vitamin B12 in your body operating. (Remember, it's recycled). Any extra and you will start to feel very unwell indeed. When the Australian lady's artificial hip was replaced with polyethylene and ceramic parts, she began to feel better in a matter of weeks.

HARD AND FAST

Transition metals don't just make good catalysts for biological reactions. They make good catalysts full stop. Nickel, a silvery metal used in the manufacture of coins and high-spec motor parts, can also drive reactions that make fats like margarine go hard. These hydrogenation reactions add hydrogen atoms to carbon-containing molecules, turning 'unsaturated' molecules (molecules with bonds to spare) into saturated ones. Around the turn of the 20th century, the French chemist Paul Sabatier realized that nickel, cobalt, iron and copper could all help hydrogenate unsaturated acetylene (C_2H_2) to ethane (C_2H_6). He started using nickel, the most effective, to hydrogenate all sorts of carbon-containing compounds. Later, in 1912, he won a Nobel Prize for his work on hydrogenation 'in the presence of finely disintegrated metals'. By this time, the food industry had adopted nickel as a catalyst for turning liquid vegetable oil into hardened margarine. Crisco, a brand of vegetable shortening used for baking, had become the first product containing the man-made fat.

Photocatalysis

Photocatalysis refers to chemical reactions that are driven by light. The idea has been employed in self-cleaning windows that break down dirt while the sun is shining. An even more space-age application is NASA's photocatalytic 'scrubbers', used by astronauts growing crops in space to break down the chemical ethylene, which causes rotting.

The trouble with the nickel process is that the catalyst also yields trans fats – partially hydrogenated contaminants that are blamed for health issues including high cholesterol and heart attacks. In the early 2000s, governments caught on to the problem and started capping the amount of trans fats in food. Today's Crisco shortening contains no trans fats.

Not all catalysts are transition metals – lots of different elements and compounds help speed up reactions. But the 2005 Nobel Prize for Chemistry was awarded for another set of reactions driven by metal catalysts – metathesis reactions, which are important in making drugs and plastics. And cobalt is now being used in cutting-edge chemistry to rip the hydrogen from water (see page 200) so that it can be used as a clean fuel.

The condensed idea
Reusable reaction drivers

13 Redox

Many common reactions are driven by the shifting around of electrons between one type of molecule and another. Rusting and photosynthesis in green plants are examples of these types of reactions. But why do we call them 'redox' reactions?

While it may sound like a sequel to an action movie, redox actually refers to a type of reaction, one that's fundamental to chemistry and to many chemical processes in nature, like photosynthesis in plants (see page 148) and the digestion of food in your gut. It's a process that often involves oxygen, which might explain the 'ox' part of redox. But to really understand why these reactions are called redox, we have to think about reactions in terms of what's happening with the electrons in them.

A lot of what goes on in chemical reactions can be attributed to the whereabouts of electrons, the negatively charged particles that form clouds around each atom's core. We already know that electrons can stick atoms together – they can be shared in the bonds that create chemical compounds (see page 20) – and that when they are lost or gained it upsets the balance of charge, resulting in positively or negatively charged particles, otherwise known as ions.

LOSS AND GAIN

Special terms are used by chemists for the loss and gain of electrons. When an atom or molecule loses electrons, this process is referred to as oxidation, while an atom or molecule that has gained electrons is said to have been reduced. There are various ways of remembering this, but perhaps the

TIMELINE

3 BILLION YEARS AGO	17TH CENTURY	1779
Photosynthesis begins with cyanobacteria	"Reduction" used to describe transformation of cinnabar (mercury sulfide) into mercury	Antoine-Laurent Lavoisier calls the component of air that reacts with metal *oxygène*, or oxygen

Oxidation states

It's all very well saying that redox reactions involve the transfer of electrons, but how do we work out where the electrons go, and how many? This requires knowing something about oxidation states. Oxidation states tell us the number of electrons that an atom can gain or lose when it partners up with another atom. Let's start with ionic compounds – with ions the clue is in the charge. The oxidation state of an iron ion (Fe^{2+}), which is missing two electrons due to oxidation, is +2. So we know it's looking to find another two electrons. Easy, right? It's the same with any ion. In table salt ($NaCl$) the oxidation state of Na^+ is +1 and the oxidation state of Cl^- is –1. What about covalently bonded compounds, such as water? In water, it's as if the oxygen atom steals two electrons from two separate hydrogen ions to fill up its outer shell, so we can consider its oxidation state as –2. Many transition metal elements, such as iron, have different oxidation states in different compounds, but often you can work out where electrons are going to go by knowing an atom's 'normal' oxidation state. This is often (but not always) defined by its position in the Periodic Table.

COMMON OXIDATION STATES:

IRON (III), ALUMINIUM	3
IRON (II), CALCIUM	+2
HYDROGEN, SODIUM, POTASSIUM	+1
INDIVIDUAL ATOMS (UNCHARGED)	0
FLUORINE AND CHLORINE	–1
OXYGEN, SULFUR	–2
NITROGEN	–3

easiest way is OIL RIG: **O**xidation **I**s **L**oss (of electrons); **R**eduction **I**s **G**ain (of electrons).

Why refer to the loss of electrons as oxidation? Surely oxidation is a reaction that involves oxygen? Well, it is sometimes, and this makes oxidation a bit of a confusing term. Rusting, for example, is a reaction between iron, oxygen and water. So, it is an oxidation reaction that involves oxygen. But it also provides an example of the other type of oxidation. During the rusting reaction, iron atoms lose electrons, forming positively charged ions. Ions of

1880	**1897**	**20TH CENTURY**	**2005**
Invention of the battery	Discovery of electrons by Joseph John Thomson	Term 'redox' used to describe reduction-oxidation reactions	Mega Rust conference established

iron, as it were. Here's how chemists would show what happens to the iron (Fe) in this reaction:

$$Fe \rightarrow Fe^{2+} + 2e^-$$

"$2e^-$" represents the two negatively charged electrons that are lost when one atom of iron is oxidized.

> **THERE ARE MORE THINGS THE MARINES SHOULD BE DOING OTHER THAN BUSTING RUST.**
> Matthew Koch, program manager for corrosion prevention and control, The US Marine Corps

These two different meanings of oxidation are in fact related – the term oxidation is expanded to include reactions that don't involve oxygen. As above, chemists describe an ion of iron in terms of how many electrons it has lost compared to its uncharged state. Losing two electrons gives it a 2^+ positive charge, with two more positively charged protons than it has negatively charged electrons.

REACTION OF TWO HALVES

What happens to the electrons? They can't just disappear – to understand where they go, we've also got to account for what's happening to the oxygen in this rusting process. At the same time as iron is losing electrons, the oxygen is gaining electrons (it's being reduced) and joining up with hydrogen to form hydroxide (OH^-) ions.

$$O_2 + 2H_2O + 4e^- \rightarrow 4OH^-$$

There is an oxidation reaction and a reduction reaction that happen simultaneously and they can be put together, like this:

$$2Fe + O_2 + 2H_2O \rightarrow 2Fe^{2+} + 4OH^-$$

When reduction and oxidation happen at the same time, it's called redox! The two 'halves' of the reaction are appropriately referred to as half-reactions.

In case you're wondering why we haven't got rust (iron oxide) yet, it's because the iron and hydroxide ions have to react with each other to form iron hydroxide (Fe(OH)$_2$), which then reacts with water and more oxygen to make hydrated iron oxide (Fe$_2$O$_3$·nH$_2$O). The redox reaction above is just part of a larger, multistage rusting process.

SO WHAT?

Knowing how rusting works is actually pretty important, because it costs industries like shipping and aerospace billions of dollars a year. The American Society of Naval Engineers holds an annual conference called 'Mega Rust' to bring together researchers who work on corrosion prevention.

Oxidizing agents and reducing agents

In a chemical reaction, a molecule that drags electrons from another molecule is called an oxidizing agent - it causes electron loss. (Remember, in OIL RIG, oxidation is loss of electrons.) So it makes sense that a reducing agent is one that donates electrons – it causes reduction or electron gain. Bleach, sodium hypochlorite (NaOCl), is a really strong oxidizing agent. It bleaches clothes by pulling electrons away from dye compounds, changing their structure and destroying their colours.

A more useful example of a redox reaction is what happens in the Haber Process (see page 68), which is important in making fertilizers, or in a simple battery. If you think about the fact that the electrical current from a battery is a stream of electrons, you may wonder where all the electrons come from. In a battery, they flow from one 'half-cell' to another – each half-cell provides the setting for a half-reaction, with one releasing electrons through oxidation and the other accepting them through reduction. In the middle of the flow of the electrons is whatever device you're trying to power.

The condensed idea
Giving and receiving electrons

14 Fermentation

From Neolithic wine to pickled cabbage, and from ancient beer to Icelandic shark-meat delicacies, the history of fermentation is entwined with the history of human food and drink production. But as archaeologists have discovered, we've been exploiting the fermentation reactions driven by microbes since before we even knew that microbes existed.

In 2000, Patrick McGovern, a chemistry graduate turned molecular archaeologist from the University of Pennsylvania, travelled to China to check out some 9,000-year-old Neolithic pottery. He wasn't interested in the pottery *per se*, rather the scum that was stuck to it. Over the next couple of years he and his US, Chinese and German colleagues subjected fragments of pottery from 16 different drinking vessels and jars found in the Henan province to various chemical tests. When they were finished, they published their results in a major scientific journal, along with findings from fragrant liquids that had remained sealed for 3,000 years inside a bronze teapot and a lidded jar in two separate tombs.

The scum provided evidence for the earliest known fermented drink, made from rice, honey and fruit from hawthorn trees or wild grapes. There were similarities between the chemical signatures of the ingredients and those of modern rice wine. As for the liquids, the team described them as filtered rice or millet 'wines', probably helped along the way to fermentation by fungi that would have broken down the sugar in the grain. McGovern has since claimed that the ancient Egyptians were brewing beer 18,000 years ago.

TIMELINE

7,000–5,500 BC	1835	1857
Early Chinese ferment drinks	Charles Cagniard de la Tour observes yeast budding in alcohol	Louis Pasteur confirms live yeast needed to produce alcohol

LIVING PROOF

Brewing is certainly an ancient tradition, but it's only the advent of modern science that has revealed how it works. In the mid-19th century, a small band of scientists formulated the 'germ theory' of disease – that diseases are caused by microbes. Just as most people didn't believe that living organisms caused disease, they didn't believe that living organisms had anything to do with the fermentation process for producing alcohol. Although yeasts had been used for years for brewing and baking, and even linked to the reactions that made alcohol, they were considered inanimate ingredients, not living organisms. But Louis Pasteur, the scientist who invented the rabies vaccine and gave his name to the pasteurization process, persisted in his studies of wine and disease.

> **[THE] FERMENT PUT INTO DRINK TO MAKE IT WORK; AND INTO BREAD TO LIGHTEN AND SWELL IT.**
>
> Definition of yeast, 1755 English dictionary

With the invention of better microscopes, views on the nature of yeast had begun to change. Finally, Pasteur's 1857 paper, *Memoire sur la fermentation alcoolique*, detailed his experiments on yeast and fermentation, and firmly established that for alcohol to be made by fermentation, yeast cells must be alive and multiplying. Fifty years later, Eduard Buchner won the Nobel Prize in Chemistry for discovering the role of enzymes (see page 132) in cells, after originally working on the enzymes that drive the booze-producing reactions in yeast.

BUBBLE AND BAKE

The reaction that we now associate with fermentation is:

$$\text{sugar} \rightarrow (\text{yeast}) \rightarrow \text{ethanol} + \text{carbon dioxide}$$

The sugar feeds the yeast and the yeast enzymes work as natural catalysts (see page 48) that drive the conversion of sugar from fruit or grains into

1907
Eduard Buchner wins Nobel Prize for work inspired by yeast-fermentation enzymes

2004
Evidence for 9,000-year-old booze published

ethanol – a type of alcohol (see Deadly drinks, below) – and carbon dioxide. The same species of yeast (*Saccharomyces cerevisiae*), but a different strain, is used in brewing. Each packet that a brewer adds contains billions of yeast cells, but there are also wild yeast growing on grain and fruits, including apple skins in cider-making. Some brewers try to cultivate these wild strains, while others avoid them because they can produce off-flavours. Both brewing and baking produce alcohol, but during baking the alcohol evaporates away.

It's the carbon dioxide by-product that gives bread its airy texture – the bubbles get trapped in the dough. Bubbles, of course, are also the key to a good glass of champagne. When vintners make sparkling wine they allow most of the bubbles to escape, but near the end of the fermentation process, they seal the bottles, trapping the bubbles and creating the pressure that will cause the cork to pop. The carbon dioxide gas trapped in a champagne bottle actually dissolves in the liquid to form carbonic acid. It's only when it escapes in the fizz that it turns back into carbon dioxide.

Deadly drinks

Chemically, an alcohol is a molecule that contains an OH group. Ethanol (C_2H_5OH) is often thought of as synonymous with pure alcohol, but there are plenty of other alcohols. Methanol (CH_3OH) is the simplest, containing just a single carbon atom. It is also known as 'wood alcohol' because it can be produced by heating wood in the absence of air. Methanol is actually a lot more toxic than ethanol and causes occasional deaths by poisoning when it is accidentally consumed in alcoholic drinks. There is no easy way for the drinker to detect it, but it is usually only produced in very small quantities in commercial brewing processes. Home-brewing and buying bootleg alcohol is more dangerous in this respect. The chemical kills because when it enters the body it is transformed into methanoic acid – or formic acid – a chemical more commonly associated with descaling products and ant bites. In 2013, three Australian men reportedly died from methanol poisoning after drinking homemade grappa. Ironically, one way of treating methanol poisoning is to drink ethanol.

Methanol

Ethanol

ALCOHOL AND ACID

Don't be fooled into thinking fermentation is something that only happens in beer and bread, or only with yeast (see Lactic acid bacteria, right). Before the refrigerator, fermentation was a useful way to preserve fish. In Iceland, dried, fermented shark meat known as kæstur hákarl is still a delicacy. It's also famous for making celebrity chef Gordon Ramsay gag. Though fermentation often means to turn sugar into alcohol it can also mean turning it into acid. The sauerkraut commonly eaten in Germany and Russia is a fermented product – cabbage that has been fermented by bacteria and preserved by pickling in the acid they produce.

Lactic acid bacteria

In yoghurt and cheese, bacteria convert milk sugar (lactose) into lactic acid. The bacteria responsible for this transformation are known as lactic acid bacteria and we have been using them for millennia to ferment our food. A similar transformation occurs in your muscles when they metabolize sugar without oxygen. The build-up of lactic acid produces the painful burning sensation in the muscles when exercising.

In recent years, fermented foodstuffs have been associated with a whole raft of health benefits. Studies have linked fermented dairy products to reduced risk of heart disease, stroke, diabetes and death. It's thought that the live microbes in fermented products affect beneficially the communities of bacteria living in your gut. Officially, however, health guidelines are more cautious and perhaps rightly so, because we still have a lot left to learn about the role of the bacteria in our insides.

So while today's health foods are a far cry from 9,000-year-old wine, they have something in common – the live microbes involved in driving the chemical reactions that create the final mouth-watering (or gag-inducing) product.

The condensed idea
The bread and booze reaction

15 Cracking

There was a time when oil was only good for burning in old-fashioned lamps. We've come a long way since then and it's all due to cracking – the chemical process that breaks up crude oil into the many useful products that fill (and pollute) our modern world, from petrol to plastic bags.

It's funny to think that our cars are powered by dead stuff. Petrol, or gasoline, is composed basically of prehistoric plants and animals that have been squashed under rocks for millions of years to make oil, and then drilled out and turned into something we can burn to produce power. The part of this process that might seem slightly mysterious to those unfamiliar with petroleum chemistry is the 'turned into something' bit.

The chemical trick that turns the dead stuff we get from under the rocks – crude oil – into useful products is called cracking. This is much more than solely fuel. Many of the things we use every day are, in fact, products of cracking. Anything made from plastic (see page 160), for instance, probably started out at an oil refinery.

A WORLD BEFORE CRACKING
In the 19th century, before cracking was invented, kerosene (see Jet fuel, page 62) was one of the only useful petroleum products. Kerosene lamps were the new, fashionable way to light your home, even if they did result in a lot of fires. The fuel itself was obtained by distilling oil – heating it up to a specific

TIMELINE

1855	1891	1912	1915
Benjamin Silliman suggests products of petroleum distillation could be valuable	Russian patent for thermal cracking granted	US patent for thermal cracking granted	The National Hydrocarbon Company became Universal Oil Products

temperature and waiting for the kerosene fraction to boil and condense. Gasoline was among the fractions which boiled off too readily and was often dumped in nearby waterways because refiners didn't know what else to do with it. The myriad possibilities within crude oil remained hidden, but not for long.

In 1855, an American chemistry professor, Benjamin Silliman, who was always being asked for his opinion on matters of mining and mineralogy, reported on the 'rock oil' of Venango County in Pennsylvania. Some of the observations he made in his report appeared to prophesize the future of the petrochemical industry. He noted that, when heated, the heavy rock oil would slowly vaporize over a matter of days, producing a succession of lighter fractions that Silliman thought could prove useful. An editor at *American Chemist* later remarked that he 'anticipated and described most of the methods which have since been adopted' in the petrochemical industry.

> ... THERE IS MUCH GROUND FOR ENCOURAGEMENT IN THE BELIEF THAT YOUR COMPANY HAVE IN THEIR POSSESSION A RAW MATERIAL FROM WHICH, BY SIMPLE AND NOT EXPENSIVE PROCESSES, THEY MAY MANUFACTURE VERY VALUABLE PRODUCTS.
>
> Benjamin Silliman, reporting to his client

WHAT'S THE CRACK?

Today, the lighter fractions like gasoline – the ones refiners were dumping in rivers – have the greatest value. What really turned rock oil into big business was the invention of cracking – first thermal cracking, then a new process involving steam and finally the development of modern catalytic cracking, driven by synthetic catalysts (see page 48).

Although the origins of cracking are not entirely clear, patents for the thermal-cracking process were granted in Russia in 1891, and in the US in 1912. The term cracking is almost a literal description of what's happening in the chemical process it describes: longer hydrocarbon chains are splitting up into smaller molecules. The cracking process enables

1920	1936	2014
First petrochemical, isopropanol, made by Standard Oil company	Exxon Mobil Oil (then Socony Vacuum Oil) and Sun Oil build catalytic crackers	Kerosene made from carbon dioxide, water and sunlight, via the Fischer–Tropsch process

products collected from straightforward distillation to be tailored to suit the refiner's requirements. Although it's possible to get gasoline – composed of molecules with five to ten carbon atoms – just by distilling oil, cracking means we can produce more of it. The kerosene fraction, for example, containing molecules with 12–16 carbon atoms, can be cracked to produce more gasoline.

Early cracking processes produced a lot of coke, a carbon residue that had to be cleaned out every couple of days. When steam cracking was invented, the addition of water dealt with the coke, but the products weren't quite ofthe quality required to make a petrol engine run smoothly. That advance came with the realization that the splitting of petroleum into its various products could be enhanced by a catalyst. Initially, chemists used clay minerals called

Jet fuel

Kerosene or paraffin is the thin oil that was used to light old-fashioned lamps. In some parts of the world, it's still used for lighting and heating, but one of its most important modern uses is in jet fuel. The components of kerosene are hydrocarbon molecules containing 12–16 carbon atoms, making them heavier than gasoline, less volatile and less flammable. This is why it's safer to burn it at home. It's not one single compound, but a mixture of different straight-chain and ring-structured hydrocarbon compounds that boil at around the same temperature. Kerosene is separated from crude oil by distillation and cracking, just like gasoline, but by comparison the gasoline fractions boil and are collected at a lower temperature. In 2014, chemists announced they had made jet fuel – kerosene – from carbon dioxide and water, using concentrated sunlight. The sunlight heated the carbon dioxide and water to produce syngas (hydrogen and carbon monoxide), which they turned into fuel via a well-known chemical route known as the Fischer–Tropsch process (see Synthetic fuels on pages 64 and 200).

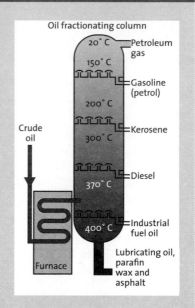

Oil fractionating column

- 20° C — Petroleum gas
- 150° C
- Gasoline (petrol)
- 200° C
- Crude oil
- Kerosene
- 300° C
- Diesel
- 370° C
- Industrial fuel oil
- 400° C
- Lubricating oil, parafin wax and asphalt
- Furnace

zeolites, which contained silicon and aluminium, until they worked out how to make artificial versions of these natural minerals in the lab.

FIGHTER FUEL

In steam cracking, the hydrocarbons often start out with single bonds and break up into shorter molecules containing double bonds. This offers up spare bonds that can be used to form new chemicals. However, with catalytic cracking, hydrocarbons aren't just broken up they are rearranged, becoming branched. Branched hydrocarbons make the best fuels, because in a combustion engine too many straight-chain molecules make the fuel 'knock' in the engine, meaning it won't run smoothly.

Just before the Second World War, the first catalytic cracker was built in Marcus Hook, Pennsylvania, giving the Allies access to fuels the German Luftwaffe didn't have. The 41 million barrels of superior jet fuel processed at the facility supposedly enhanced the manoeuvrability of the Allies' fighter planes, giving them an edge in the air.

The Shukhov Tower

On Shabolovka Street, Moscow, stands an intricately designed, 160-m-high, radio tower conceived and constructed by Vladimir Shukhov in the 1920s. Shukhov was a remarkable person, building Russia's first and second oil pipelines, as well as having a hand in the design of Moscow's water supply. He is credited with an early patent for thermal cracking – before the process was patented by Russia's great rivals, the Americans. In 2014, the Shukhov Tower narrowly escaped demolition.

While the catalytic process makes excellent fuels, it's also key to the chemical industry, producing many of the basic structures that are used to build globally important chemicals, like polythene. If oil ever runs out, we'll need to have come up with alternative ways of producing these products. Manufacturers are already turning to living plants instead of long-dead ones to make chemicals. One German company is selling paint made from mignonette, a sweet-smelling plant used in perfumes.

The condensed idea
Making oil work for us

16 Chemical synthesis

How many of the products you use in your home every day contain synthetic – man-made – compounds? While you might realize that medicines and additives in many of the foods you eat are products of the chemical industry, you might not have thought about your stretchy underwear or the stuffing in your sofa.

Think about everything you're wearing right now. Do you have any idea what your shirt or underwear are made of? Check the labels: what's viscose? Where does elastane come from? Now check your bathroom cabinet. What are the ingredients in your toothpaste? Your shampoo? What about propylene glycol? It gets even more bewildering when you start opening up kitchen cupboards, digging out boxes of medicine (see page 176) and studying the ingredients on the back of a packet of chewing gum.

It's incredible to think that so many of the chemicals that go into our clothes, food, cleaning products and medicines have been developed by chemists just in the last century. These synthetic chemicals were invented in a lab and are now manufactured on an industrial scale.

> **I AM JUST A GUY WEARING SPANDEX THAT TURNS LEFT REALLY FAST.**
> Olympic gold medal-winning speed skater, Olivier Jean

NATURAL VERSUS SYNTHETIC

Viscose, or rayon, was the first synthetic fibre produced by chemists. Its fibres form a soft cotton-like fabric that easily absorbs dyes, not to mention sweat. An early process for making it was invented at

TIMELINE

1856	1891	1905	1925
Discovery of first synthetic dye by 18-year-old chemist, William Henry Perkin	A process for making viscose, once known as artificial silk	The first commercial viscose process	Patent granted for Fischer–Tropsch process

the end of the 19th century. Actually, viscose is not very different to a natural compound common to all plants – cellulose – but you can't just grow viscose in a field. The cellulose comes from smashed up wood, to which various chemical and physical processes are applied that transform it into crumbs of yellow cellulose xanthate. The xanthate is decomposed by acid during manufacture, leaving fibres like those of natural cotton, which is almost pure cellulose. Viscose and cotton are often blended together in fabrics.

Any process that involves exploiting chemical reactions to make some specific, useful product can be referred to as chemical synthesis. Natural products like cellulose are also made by chemical reactions – in this case, exploited by plants – but chemists tend to think of them as products of biosynthesis (see page 144) instead.

Sometimes, the chemicals that chemists make synthetically are actually copies of compounds produced in nature. In these cases, it's often about making the product cheaper or in larger quantities, rather than making something that works better than the natural product. After all, nature usually does a pretty good job. The basis of the chemical compound in the influenza drug, Tamiflu, for instance, is shikimic acid, made in the seeds of the plant from which we get the Chinese spice, star anise. But because the supply of star anise is limited, there is an ongoing effort by chemists

Synthetic fuels

Fischer–Tropsch synthesis is a process for making a synthetic fuel from various reactions of hydrogen and carbon monoxide. The two gases (together known as 'syngas') are usually produced by turning coal into a gas. This makes it possible to create the liquid fuels that we would normally associate with petroleum (see page 156) without relying on oil. In South Africa, SASOL has been producing 'synfuels' from coal for decades.

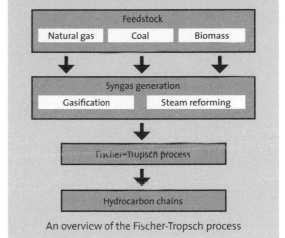

An overview of the Fischer-Tropsch process

The synthesis machine

Imagine if chemists didn't have to go through the rigmarole of designing a series of reactions to make the molecule they want. Imagine if they could just plug the identity of that molecule into a machine and the machine would decide on the best way to make it, and then go right ahead and do it. What a revolution this would be to the design of drugs and new materials. For DNA, at least, that machine already exists. DNA synthesis machines can churn out short stretches of DNA of any desired sequence. Doing the same thing for any molecule is obviously going to present a bigger challenge, not least in terms of computing power. A synthesis machine would need to devise its synthetic routes based on scanning through millions of different reactions at lightning speed and comparing billions of possible pathways. Despite scepticism, serious efforts are underway. For example, a team of British researchers working on the 'Dial-A-Molecule' project has set itself the grand challenge of making the synthesis of any molecule 'as easy as dialling a number'. Another American project has built a 'chemical Google' that knows 86,000 chemical rules and uses algorithms to find the best synthetic route.

to make the drug from scratch. Various different methods have been reported, but each must be evaluated against the cost of extracting the starting ingredients from seeds.

STRETCHY PANTS

Other synthetic products have nothing to do with nature. In fact, their 'unnatural' properties are exactly what makes them useful to us. Elastane is a shining example. You might know it better as Lycra or spandex – the stretchy, skin-tight fabric beloved by cyclists. The clothing giant Gap mixes together spandex and nylon to make its yoga-wear, while Under Armour StudioLux is a combination of spandex and polyester. Today, we're unfazed by all these fancy-sounding fibres, but the influx of spandex into the clothing market in the 1960s was a revolution.

Like the cellulose molecules in cotton fibres, the long-chain molecules in spandex are polymers containing the same chemical blocks repeated over and over again. Making the polyurethane building blocks requires one set of chemical reactions, while joining them requires another. That's perhaps why it took DuPont scientists a couple of decades to work out a viable manufacturing process. Unlike cotton fibre, the resulting 'Fiber K' – as it was initially dubbed – had some astonishing and valuable properties. Spandex fibres can expand up to six times their original length and snap back into shape. They are more durable and can withstand greater tension than natural rubber. DuPont had got itself a blockbuster product, and ladies' support garments had suddenly become a lot more comfortable.

CHEMICAL BACKBONES

Now think back to your wardrobe, your bathroom cabinet and your kitchen cupboards. Think how many of the other products you buy contain materials or ingredients that are the result of years or decades of tireless research by chemists. The number of chemical reactions required to fill your home with stuff is mind-boggling.

Many products of chemical synthesis rely on the cracking (see page 60) of oil as a reliable source of useful chemicals. If you're still wondering what polypropylene glycol is, it's the ingredient in shampoo that helps your hair absorb moisture to keep it soft, and it's made using propylene oxide – created by a reaction between the cracked chemical propylene and chlorine. Propylene oxide is also used in the manufacture of antifreeze and foams for furniture and mattresses. So while you might never have heard of it, the annual global demand for propylene oxide is over six million tons, not because it is particularly useful on its own but because it is possible to use it to make lots of different everyday products, through chemical synthesis.

In the same way, many other compounds form the chemical bones to which the meat of each industrial product is added. From drugs to dyes, plastics to pesticides, soaps to solvents – you name it, the chemical industry probably had a hand in making it.

The condensed idea
Making useful chemicals

17

The Haber Process

Fritz Haber's discovery of a cheap process for making ammonia was one of the most significant breakthroughs of the 20th century. Ammonia is used to produce fertilizers, which have helped to feed billions of people, but it is also a source of explosives – a fact not lost on those who were commercializing the Haber Process just as a world war was breaking out.

Henri Louis was the son of the engineer Louis Le Châtelier. His father, who was interested in steam trains and steel production, invited many prominent scientists around to their house. As a small boy growing up in Paris in the 1850s, Henri Louis was introduced to many famous French chemists. Some of their influence must have rubbed off, because he went on to become one of the most famous chemists of all time, giving his name to a guiding law of chemistry known as Le Châtelier's Principle (see page 36).

Le Châtelier's Principle describes what happens in reversible reactions. Ironically, though, in attempting to perform one of the most important reversible reactions on the planet (see The ammonia-making reaction, opposite), Le Châtelier slipped up. He fluffed the experiment that would have allowed him to make the chemical now at the centre of two global industries: the fertilizer industry and the arms industry.

NITRATE WARS
Fertilizer is sometimes described as containing 'reactive nitrogen', because the nitrogen is in a form that can be taken up by plants and animals to use in making proteins. That's as opposed to all the unreactive nitrogen (N_2) floating

around in Earth's atmosphere. By the early 20th century, the world had recognized the potential of reactive nitrogen for fertilizer and begun importing the natural mineral saltpetre or potassium nitrate ($NaNO_3$) from South America, to boost crop production. A war over nitrogen-rich lands ensued and was won by Chile.

Meanwhile, in Europe, there was an urgent need to secure a plentiful source of ammonia on home turf. Turning ordinary N_2 into reactive forms like ammonia (NH_3) – 'fixing' nitrogen – was energy-intensive and expensive. In France, Le Châtelier approached the problem by taking its two components – nitrogen and hydrogen – and reacting them under high-pressure. His experiment exploded and he narrowly avoided killing his assistant.

The ammonia-making reaction

The reversible reaction for making ammonia is:

$$N_2 + 3H_2 \rightleftharpoons 2NH_3$$

It's a redox reaction (see Redox, page 52). It's also an exothermic reaction, meaning it loses energy to its surroundings and doesn't need a lot of heat to get it going. It can amble along quite happily at low temperatures. However, producing industrial quantities of ammonia does require heat. Although a higher temperature nudges the equilibrium (see Equilibrium, page 36) slightly to the left, favouring nitrogen and hydrogen, the reaction proceeds far quicker, meaning more ammonia can be made in a shorter time.

Sometime later, Le Châtelier discovered that his set-up had allowed oxygen from the air to get into the mix. He'd come pretty close to synthesizing ammonia, but it is a German scientist, Fritz Haber, whose name we now associate with the ammonia-making reaction. By the start of the First World War, ammonia had become important for another reason: it could be used to make nitroglycerine and trinitrotoluene (TNT) explosives. The ammonia that Europe had wanted as fertilizer was soon being consumed by the war effort.

THE HABER PROCESS

Had it not been for the near-death explosion, Le Châtelier might never have abandoned his work on ammonia. The Haber Process – as it became known

1909	1914	1915	1918
Fritz Haber makes ammonia in the lab	The First World War begins in Europe	Haber directs chlorine gas attack at Ypres	Haber wins Nobel Prize in Chemistry

– even made use of Le Châtelier's own theories. The important reaction in ammonia synthesis forms an equilibrium between the two reactants (nitrogen and hydrogen) and the product (ammonia). As predicted by the Le Châtelier Principle, removing some of the product upsets the status quo and encourages the equilibrium to work to reestablish itself. So in the Haber Process, ammonia is constantly removed to drive further production of ammonia.

Haber used an iron-oxide catalyst to speed up his reaction. Here again, Le Châtelier was evidently not too far off the mark. In a book published in 1936, he wrote that he tried to use metallic iron. Haber was also inspired by the work of thermodynamics theorist Walther Nernst, who had already produced

Naturally fixed nitrogen

Saltpetre is a naturally occurring mineral containing nitrogen in a reactive or 'fixed' form. Before the advent of the Haber Process, another major source of reactive nitrogen was Peruvian guano – bird droppings collected from seabirds nesting along the coast of Peru. In the late 19th century, Europe was importing both as fertilizer. There are other ways that nitrogen can be fixed. Lightning strikes convert small quantities of nitrogen in the air into ammonia. Early processes for making ammonia mimicked this process using electricity, but they were too expensive.
Certain bacteria living in nodules on leguminous plants, such as clovers, peas and beans, also fix nitrogen. For this reason, farmers often practise crop rotation to replace nutrients taken from the soil and make it more fertile for the next year's crops. Planting sweet clover gives them a 'nitrogen credit', meaning they don't need to apply as much fertilizer the next year when planting grain.

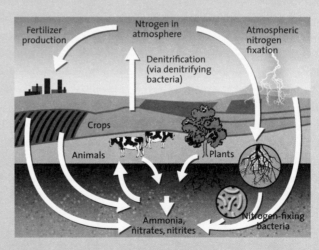

ammonia in 1907. But it was Haber who would be rewarded for his efforts. After he coaxed the first drops of ammonia from a tabletop experiment in 1909, his colleague Carl Bosch helped commercialize the process (sometimes known as the Haber–Bosch Process.) Nearly a decade later, Haber would be awarded the Nobel Prize in Chemistry, but it was to be a controversial decision.

> **I LET THE DISCOVERY OF THE AMMONIA SYNTHESIS SLIP THROUGH MY HANDS. IT WAS THE GREATEST BLUNDER OF MY SCIENTIFIC CAREER.**
> Henri Louis Le Châtelier

The nitrogen used to make fertilizers is said to have doubled crop production. In the century following Haber's discovery, as many as four billion people were fed by crops grown as a result of the cheaper, more energy-efficient way of making ammonia, hailed as 'bread from air'. But while Le Châtelier may have desperately wished that ammonia synthesis had been his discovery, he did at least manage to preserve his reputation. There were more than 100 million deaths in armed conflict during the 20th century and the Haber Process had a hand in most of them.

Haber didn't exactly do himself any favours. He went on to mastermind a chlorine gas attack that killed thousands of French troops at Ypres in April 1915. His wife, who had begged with him to give up his work on chemical weapons, shot and killed herself only days later. Haber may have won a Nobel Prize, but he isn't exactly remembered fondly. Le Châtelier, meanwhile, is remembered for his more noble efforts in explaining the principles that govern chemical equilibrium.

Ammonia is still produced in vast quantities. In 2012, over 16 billion kilograms were made in the USA alone. Scientists are still working to understand the impact of all that reactive nitrogen flowing from farms into rivers and lakes.

The condensed idea
Chemistry that dictates life and death

18 Chirality

Two molecules can look almost identical but act completely differently. This curious quirk of chemistry is all down to chirality – the idea that some molecules have mirror images, or left- and right-handed versions. The implication is that for every chiral chemical, there is one version that does the job it's supposed to do and another that does something else entirely.

Place your hands together as if you are praying – not to offer a prayer but so you can appreciate the asymmetry of your hands. Your left hand is a mirror image of your right – you might think of them as exactly the same, but they're actually exact opposites. However hard you try, you couldn't arrange your left hand to match your right. Even if modern medicine could perform perfect hand transplants, you couldn't swap them and have them do the same job.

Some molecules are like hands. They have mirror versions that can't be superimposed on one another. All the same atoms are present and, superficially, the structures look the same. But one version is a reflection of the other. The technical term for these left- and right-handed versions is enantiomers. Any molecule that has enantiomers is said to be chiral.

Any left-handed person who has tried to use a pair of right-handed scissors might have an inkling of the importance of this fact. The difference between two enantiomers of a molecule can be the difference between a chemical doing the job it was intended to do ... and not. Fuels, pesticides, drugs and even the proteins in your body are all chiral molecules.

TIMELINE

1848	1957	1961	1980
Louis Pasteur discovers chirality in sodium ammonium tartrate	Thalidomide launched – in Germany first	Thalidomide begins to be removed from sale	Term 'EPC-synthesis' (enantiomerically pure compound) is introduced

GOOD AND BAD VERSIONS

There is an entire branch of chemistry devoted to making chiral chemicals that have the correct 'handedness'. Ultimately, the aim of producing a commercial chemical is to produce enough to make a profit. So if the reactions for, say, a new drug result in a mix of left- and right-handed molecules, but only the left-handed version works, the reaction can be further optimized.

More than half of the drugs made today are chiral compounds. Although many are produced and sold as mixtures containing both enantiomers, one enantiomer usually works better. Beta blockers, which are used to treat high blood pressure and heart conditions, are a prime example. In certain cases, however, the 'wrong' enantiomer can actually be harmful.

There can be no more shocking example of a 'bad' enantiomer than thalidomide, a drug notorious for its effects on babies in the womb. The drug, which was initially prescribed as a sedative when it was launched in the 1950s, was soon being given to pregnant women to help them cope with morning sickness. Unfortunately, its mirror version was a compound that caused birth defects. It's thought that more than ten thousand babies were born with disabilities due to the effects of thalidomide. Today, there are still ongoing legal battles between the manufacturers and those who were disabled by the drug.

Racemic mixtures

Mixtures that are made up of left-handed and right-handed molecules in about equal quantities are called racemic mixtures. Sometimes these mixtures are referred to as 'racemates'. So when some of the molecules of the drug thalidomide switch from their single enantiomer, forming a mixture, they are said to 'racemize'

MAKING MIRROR IMAGES

Attempts to make thalidomide containing only the good version have failed, because it is able to switch between enantiomers inside the body (see Racemic mixtures, above), resulting in a mixture of good and bad versions.

2001

Nobel Prize in
Chemistry awarded
for asymmetric synthesis
of drugs

2012

Analysis of meteorite fragments
from Tagish Lake in Canada finds
excess of left-handed amino acids

How do you tell if a compound is chiral?

Two molecules that are made up of the same atoms – but arranged in a different order – are called isomers. But in chiral compounds, the two isomers have all their atoms attached in the same order. They're practically identical in every way, except that they are reflections of each other. So how can you look at a molecule and tell whether it's chiral or not? The way to tell is that a chiral molecule has no plane of symmetry. So if you can draw an imaginary line down the molecule's centre and match its two sides – something like a paper snowflake cut-out – then it's not chiral. Remember, however, that molecules are 3-D objects, so it's not always as easy as it seems to draw a line down the middle. It can, in fact, be very difficult to tell whether a molecule is chiral simply by looking at its structure on paper. For complex molecules, it can help to build a 3-D model using sticks and bits of modelling clay. (See also Sugars and stereoisomers, page 137.)

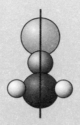

Plane of symetry

No plane of symmetry anywhere in this molecule

Some compounds can be separated from their chiral counterparts, and it is also possible to fix reactions so that the products contain only a single enantiomer. In 2001, two American chemists and one Japanese chemist shared the Nobel Prize in Chemistry for their work on chiral catalysts – which they used to make chiral compounds including drugs. The prize was partly awarded to William Knowles for designing reactions that produced only the 'good' version of a Parkinson's drug, called dopa. Like thalidomide, its enantiomer is toxic.

In recent decades, drug approval authorities have become more aware of potential problems with enantiomers. Drug companies used to produce drugs containing mixtures of left- and right-handed molecules, regarding less effective mirror versions as unwanted baggage. Now, they try to make drugs containing only single enantiomers.

LIFE IS SINGLE-HANDED

Nature, however, does things differently. When chemists make chiral compounds in the laboratory they often form in approximately equal quantities of left- and right-handed molecules. But biological molecules follow a predictable pattern of handedness. Most notably, amino acids, which are the building blocks of proteins, are left-handed, while sugars are right-handed. No one knows exactly why this is, although researchers who study the origins of life on Earth have different theories about it.

Some scientists have suggested that molecules brought to early Earth by meteorites could have given life on Earth a nudge in the right or left direction. Meteorites have been known to crash to Earth carrying amino acids, so it's possible that any slight excess in left-handed meteorite molecules was taken up by organic compounds present in the primordial seas, just as the molecules of life were forming here on Earth. Whatever happened, it seems likely there were some initial imbalances in left- and right-handed molecules that became magnified as time went by. To be sure, we can't go back in time to check this theory, so we can't say for sure that the observed single-handedness didn't develop later, when life had already become more complex.

> **CHIRALITY HELD ALICE'S ATTENTION AS SHE PONDERED THE MACROSCOPIC WORLD SHE GLIMPSED THROUGH THE LOOKING GLASS ...**
> Donna Blackmond

Chirality in biological molecules is not just a curiosity. It leads us back to our understanding of synthetic chiral compounds and their actions as drugs. Drugs work by interacting with the biological molecules in our bodies. For a drug to have any effect, it first has to 'fit'. Think of it like a hand slipping into a glove – only the left glove slides easily onto the left hand.

The condensed idea
Mirror molecules

19 Green chemistry

The last few decades have seen the rise of green chemistry – a more sustainable approach to doing science that reduces waste and encourages chemists to be cleverer about how they design their reactions. And it all started when bulldozers arrived in a backyard in Quincy, Massachusetts.

Paul Anastas grew up in Quincy, Massachusetts, USA, where his parents' home – at one time – offered a view of the Quincy wetlands. That view was destroyed by big business and big glass buildings, inspiring Anastas to write an essay about the wetlands that earned him an Award of Excellence from the president, aged nine. Nearly two decades later, after getting his PhD in organic chemistry, he started work at the United States Environmental Protection Agency (EPA), and it was there that he wrote his manifesto for a smarter, greener, new kind of chemistry. He would later become known to chemists everywhere as the 'father of green chemistry'.

At just 28 years old, Anastas's concept of 'green chemistry' was to reduce the environmental impact of chemicals, chemical processes and industrial chemistry. How? Basically, by finding cleverer, more environmentally friendly ways of doing science, by cutting down on waste and by reducing the amount of energy chemical processes needed to consume. It was a concept he knew might not go down well with industry, so he sold it on the basis that working smarter should also mean working cheaper.

TIMELINE

1991	1995	1998
Paul Anastas coins the term 'green chemistry'	Presidential Green Chemistry Challenge set up	Anastas and John Warner publish *Green Chemistry: Theory and Practice*

Greener desalination

Population growth and drought mean that water is becoming scarcer. Lots of cities around the world have desalination plants so they can source some of their drinking water by extracting the salt from seawater. But getting the salt out is an energy-intensive process based on forcing the water through a thin membrane containing tiny holes. The technique is called reverse osmosis. Making the specialized membranes used in reverse osmosis often involves a lot of chemicals, including solvents. In 2011, one of the winners of the Presidential Green Challenge awards was a company that developed a way to make new, inexpensive polymer membranes, which could be made using fewer harmful chemicals. Kraton's NEXAR membranes are also intended to save energy in desalination plants and could potentially slash energy costs by half.

Applying pressure greater than the osmotic pressure drives desalination of seawater

Pressure

Semi-permeable membrane

Fresh water

Water

Seawater

Osmotic pressure

THE 12 PRINCIPLES OF GREEN CHEMISTRY

In 1998, together with Polaroid chemist John Warner, Anastas laid down his 12 Principles of Green Chemistry. In essence they were.

1. Produce as little waste as possible
2. Design chemical processes that make use of every atom you put in
3. Don't use hazardous reactants; don't make hazardous by-products
4. Develop new products that are less toxic
5. Use safer solvents and less of them
6. Be energy efficient

2011

Market for green chemistry reaches $2.8 billion

2020

Market for green chemistry predicted to reach $98.5 billion

7.   Use raw materials that can be replaced
8.   Design reactions that produce only the chemicals you need
9.   Make use of catalysts to increase efficiency
10. Design products that degrade safely in the environment
11. Monitor reactions to avoid waste and hazardous by-products
12. Choose approaches to minimize accidents, fires and explosions

> **WE'LL KNOW THAT GREEN CHEMISTRY IS SUCCESSFUL WHEN THE TERM 'GREEN CHEMISTRY' DISAPPEARS BECAUSE IT'S SIMPLY THE WAY WE DO CHEMISTRY.**
>
> Paul Anastas, quoted in *The New York Times*

The 12 Principles were all about being more efficient with what you used, and what you created, and emphasizing chemicals that were less harmful to people and the environment. Common sense, you might think. But for a chemical industry that had been doing things in a very different way for a very long time, it needed spelling out.

AT THE PRESIDENT'S HOUSE

Anastas had gone swiftly from lowly chemist to section chief and from there to Director of a new Green Chemistry Program at the EPA. In his first year as director, he'd proposed a set of awards to honour achievement in the field of green chemistry – achievement by both academic scientists and companies. The president himself, Bill Clinton, endorsed the awards as the Presidential Green Chemistry Challenge. They are still going strong.

In 2012, one of the winners was a company called Buckman International, whose chemists had come up with a way to make stronger recycled paper without wasting lots of chemicals and energy. Taking note of article nine on Anastas and Warner's list, they adopted enzymes – biological catalysts – to shepherd reactions that produce wood fibres with just the right structure. They reckoned enzymes could save a single papermaking plant $1 million a year, supporting the theory that working smarter means working cheaper.

Other awards have been handed out for green ways to make cosmetics, fuels and membranes that purify salt water. Anastas, meanwhile, was soon snapped up by Clinton himself and began working at the White House Office of Science and Technology on environmental policy. He'd gone from presidential award winner at age nine, to setting up his own presidential award, to working at the White House, and was still only 37.

GREEN FUTURE

According to the EPA's own figures, the amount of hazardous chemical waste produced in the US fell from 278 million tonnes in 1991 – when Anastas coined the term green chemistry – to 35 million tonnes in 2009. Companies were starting to pay more attention to their impact on the environment. Let's not get carried away though – Anastas has done very well for himself, come up with some great ideas and reached the White House, but the industry's problems were not solved at a stroke. Far from it. Many important chemicals that form the basis of everyday products are still made by refining oil, which is not a renewable resource and can be very polluting. There is a lot more than can be done.

Green chemistry is still a young field. It's expected to grow rapidly by some estimates to nearly $100 billion by the end of the decade. But Anastas won't be happy until he's painted the entire chemical industry green. In an interview with the leading science journal *Nature* in 2011, 20 years down the line, Anastas said his ultimate goal for chemistry is for it to completely adopt the principles of green chemistry. When that goal is achieved, the term 'green chemistry' will cease to exist altogether – green chemistry will just be chemistry.

Atom economy

The principles of green chemistry refer to a concept called 'atom economy', which wasn't developed by Anastas and Warner, but by Stanford University's Barry Trost. For any reaction, you can estimate the total number of atoms in the reactants and compare it to the total (estimated) number of atoms in the products. This ratio tells you how economical you've been in your use of atoms. In green chemistry, every atom counts.

The condensed idea
Chemistry that doesn't hurt the environment

20 Separation

Whether it's separating coffee granules from our morning brew, the scent of jasmine from its flowers or heroin from blood at a crime scene, there are few techniques more useful in chemistry than those for separating one substance from another. In Dutch, chemistry translates as 'the art of separation'.

In every TV detective show, there's a part where the forensics team rocks up and takes over the crime scene. We don't see what they do. We don't know what goes on back at the lab. All we know is they arrive in their paper-thin, disposable crime-scene suits and then a few minutes later, Detective Inspector Whodunnit is reading the results off a sheet of paper. Crime solved.

It would be interesting to see what work was actually done in the forensics department. One of the things that the forensics guys are experts at is chemical separations. Imagine they get back from a particularly nasty crime scene. Blood spattered everywhere and evidence of drug-taking. One of the things they want to do is establish who was taking what drugs. They have blood samples but how do they get the drugs out of them so they can work out what they are? The problem they're dealing with is a much more complicated version of picking paperclips out of a bowl of rice. In this case the two substances are wet and can't be separated by hand.

CHROMATOGRAPHY
What the forensics guys are going to use, invariably, is some sort of chromatographic technique. Essentially, they'll try to get the drug to stick to something, the idea being that the drug is attracted to whatever the

TIMELINE

Ancient Egypt	1906	1941	1945
Scent extracted from flowers using fat	First published paper on chromatographic techniques	Martin and Synge invent partition chromatography	Evika Cremer and Fritz Prior develop gas chromatography

'sticky' material is, while the blood runs straight off. It's a bit like using a magnet to pull the paperclips out of the bowl of rice. In forensics terminology, the drug, or the paperclip, is the analyte – the thing that's being analysed.

PERFUME AND PAINT

In principle, any modern chromatography is pretty similar to extraction techniques that have been used for centuries in industries like perfume-making. The sticky material doesn't have to be solid. When perfumers extract the scent of jasmine from jasmine flowers, for instance, they use liquid chemicals like hexane. The important point is that the scent compounds have more of an affinity for the liquid than other compounds in the flowers.

Most of us are familiar with chromatography because at school we were given pieces of paper to separate different coloured inks or pigments – our analytes. Two different pigments will have different interactions with the paper and so end up forming separate, different-coloured spots. The term 'chromatography' itself literally means 'to write with colour'. One of the first scientists to work with

Electrophoresis

Electrophoresis covers a range of methods used for separating molecules such as proteins and DNA using electricity. Samples are added to a gel or fluid and molecules separate according to their surface charge - negatively charged molecules move towards the positive electrode, whilst positively charged molecules move towards the negatively electrode. Smaller molecules travel faster because they face less resistance, so the components are also separated by size.

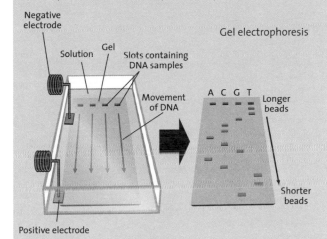

1952
Martin and Synge awarded Nobel Prize for Chemistry

1970
Csaba Horváth coins HPLC – first high-pressure, then high-performance liquid chromatography

1990
First report on capillary electrophoresis in DNA sequencing

Sorting the wheat from the flour

Separation methods are common in the food analysis industry. There are companies that help food manufacturers track down chemicals and other foreign bodies that have got into their products and finding them involves separating them from the other ingredients. One problem is contamination of products that are sold as gluten, wheat or lactose-free. Even tiny amounts of the offending molecules can cause illness in people with sensitivities. Food analysts can use chromatographic techniques to find the impurities. For example, a 2015 study by German chemists described a new method for identifying wheat contaminants in spelt flour. The trouble with these two grains is that they are often crossed together to make wheat/spelt hybrids. Spelt is generally easier to digest, but hybrids contain genes from both and make plenty of the same proteins. However, the researchers were able to identify a gliadin protein that was unique to wheat. They showed that it would be possible to carry out high-performance liquid chromatography (HPLC) on a spelt flour to determine whether it had wheat in it – the gliadin protein contaminant would be noticeable in the pattern on the chromatogram. The same technique could also be used to classify different crops by their wheatlike and speltlike proteins.

chromatographic techniques back in the 1900s was a botanist who used paper to separate coloured plant pigments. It wasn't until 1941, however, that Archer Martin and Richard Synge combined liquid–liquid extraction methods, like those used in perfumery and chromatography, and invented modern 'partition chromatography', using a gel to separate amino acids.

Now, while it is true that chromatography has certain similarities to extraction, chromatography is more likely to be used by our forensic scientists. That's because it is better able to separate the small quantities of chemicals – drugs, explosives, fire residues or other analytes.

MOVING ON UP

In the school paint experiment, there's what's called a stationary phase, which is the paper (the 'magnet' or sticky material) and a mobile phase, which is the paint, because it moves up the paper. Although today's forensics labs are more hi-tech, these phases are still given the same names. Two very widely used techniques are gas chromatography and high-performance liquid chromatography (HPLC), which uses high pressures. Both of these will separate drugs, explosives and fire residues. They can even be coupled directly to mass spectrometers (see page 84) that can help forensics teams identify the exact chemicals in question. The molecular 'signature' of the analyte might be recognizable as heroin, for example.

To confirm the identity of the person with heroin in their blood, forensic scientists can also use capillary electrophoresis (see Electrophoresis, below), another common separation technique. Here, electricity forces DNA (the analyte) to move through tiny channels, separating into a different pattern depending on the person's DNA profile. The profile or 'DNA fingerprint' can be checked against a reference sample, for example, from blood or hair. The real skill of the forensic scientist lies in deciding which techniques to use and how best to combine them. The end result may be detecting heroin, but it could take several separation steps to get to a point where the drug could be detected at all.

OTHER SEPARATION TECHNIQUES

Of course, forensic scientists aren't the only ones using separation techniques, even if they appear to be the most glamorous. Separations are standard analytical methods. Others that deserve a mention are good old distillation, which separates liquids based on their boiling points (see page 60), and centrifugation, which uses a centrifuge machine to spin and separate particles based on their different densities. You may be starting to see a pattern here; all chemical separations work simply by taking advantage of different properties of the chemicals they try to separate. For one final example, think about a paper coffee filter, which physically separates solid coffee granules from liquid coffee – a separation based on states. Filtering is also a common technique in chemistry labs, although chemists may use vacuums and pumps to help the process along. There are other laboratory methods, that tell chemists the constituents of mixtures and compounds.

> **EVEN TODAY, IN HOLLAND, CHEMISTRY IS CALLED 'SCHEIKUNDE', OR 'THE ART OF SEPARATION'.**
>
> Professor A. Tiselius, member of the Nobel Committee for Chemistry (1952)

The condensed idea
What detective dramas won't teach you

21 Spectra

To most of us, spectra are baffling spiky or lumpy graphs that appear in the results sections of scientific papers. But to trained observers, these patterns reveal the intricate details of a compound's molecular structure. One of the methods used to create these images is also the basis of a key technique in cancer diagnosis and treatment – the MRI scan.

When someone with a brain tumour goes for an MRI – magnetic resonance imaging – scan, they're asked to lie inside a machine containing a very powerful magnet while it creates an image of their brain. That image will be able to distinguish the tumour from the surrounding tissues and will be used to inform a doctor's decision about whether and how to operate. Effectively, the MRI machine gets inside the patient's head without ever causing them any pain or doing any internal damage. All they have to do is lie very still so as not to disturb the image.

The fact that MRI is harmless is something that often has to be emphasized. One reason is that it is directly descended from nuclear magnetic resonance (NMR), and anything associated with the word 'nuclear' understandably makes people jumpy. Both MRI and NMR work, based on a natural property of certain atoms: their nuclei act like tiny magnets. When a powerful magnetic field is applied, it affects the behaviour of the nuclei. By tuning into this behaviour using radio waves, an NMR machine is able to extract information about the environment of the nuclei and an MRI machine is able to extract information about a patient's brain.

TIMELINE

1945	1955	1960
Edward Purcell and Felix Bloch both independently discover NMR phenomenon	William Dauben and Elias Corey use NMR to discover molecular structures	First commercially successful NMR – the Varian A-60

FROM NMR TO MRI

Paul Lauterbur, the chemist who had such an instrumental role in the development of MRI – and was awarded a Nobel Prize in 2003 for his efforts – was originally an NMR expert. He learned the technique at the Mellon Institute Laboratories in the 1950s while studying for his PhD and carried on working with it during a brief spell in the US Army. He was supposedly the only person who knew how to operate the Army Chemical Center's

Newborn testing

Mass spectrometry is one of the techniques used to analyse chemicals in the blood of newborn babies – it can identify molecules that might indicate inherited diseases. For example, high levels of an amino acid called citrulline suggest that a baby may have an inherited disease called citrullinemia, which causes toxins to accumulate in the blood and can lead to vomiting, seizures and growth suppression. Involved with metabolic processes, citrulline is also a useful biomarker for rheumatoid arthritis. Citrullinemia is rare but can quickly become life-threatening if it is not treated early on. Mass spectrometry is a very quick and accurate method for analysing samples. It can also be used to simultaneously detect several different compounds at once, so the same sample can be used to test for a number of different diseases.

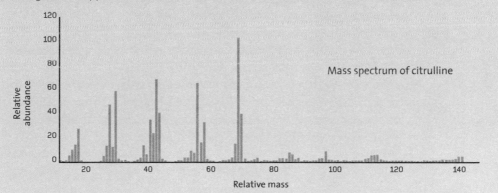

Mass spectrum of citrulline

1973	2003	2011
Paul Lauterbur introduces MRI	Nobel Prize awarded for discovery of MRI	American Chemical Society designates the Varian A-60 a National Historic Chemical Landmark

new NMR machine. It was around that time that the first commercial NMR machine – the Varian A-60 – was developed by Varian Associates. It would soon find wider use in medicine.

> **PRIOR TO THE USE OF NMR ... [A CHEMIST] COULD SPEND LITERALLY MONTHS AND YEARS TRYING TO DETERMINE THE STRUCTURE OF A MOLECULE.**
> Paul Dirac, 1963

The element most often used to produce NMR spectra was hydrogen, which is present in water – and therefore also in blood plasma and body cells. By using hydrogen nuclei as magnets, NMR can image a patient's head. In 1971, Lauterbur was alerted to some interesting research on tumour cells carried out by a medical doctor. The water content of a tumour cell is different to that of an ordinary cell and Raymond Damadian had shown that NMR could distinguish between the two – though he did his research in rats and had to sacrifice animals to get his spectra. Lauterbur not only found a way to turn the data into an (initially fuzzy) image, but found a way to do it without touching a hair on the patient's head.

By the time Lauterbur got his Nobel Prize in 2003, NMR had been around for over half a century and had become one of the most important analytical techniques employed in chemistry labs around the world. Hydrogen is a common atom in organic compounds, and in a NMR spectrum, protons show characteristic peaks that correspond to the hydrogen nuclei in different environments – in relation to the other atoms in a molecule. Plotting the positions of the hydrogen atoms in a compound can tell an organic chemist a lot about its structure – it can be used to analyse the structures of new compounds or identify those that are already known.

READING THE PEAKS

A molecule's NMR spectrum forms a pattern, a chemical fingerprint that points to its identity. But there are other kinds of chemical fingerprints and like NMR their interpretation relies on the recognition of characteristic waves or peaks within a spectrum. In mass spectrometry, the different peaks relate to different molecular fragments – ions – that are produced when molecules are blown apart by a high-energy beam of electrons. The position of the peak along a scale shows the mass or weight of individual fragments corresponding to that peak, while the height of the peak indicates the number of fragments. This allows the researcher to identify the components in an unknown substance and, by working out how the fragments fit together, the structure of the molecule.

INFRARED ANALYSIS

Another important analytical technique is infrared (IR) spectroscopy, which uses infrared radiation to make the bonds between atoms in a molecule vibrate more vigorously. Different chemical bonds vibrate in different ways and an infrared spectrum shows a range of peaks that relate to different bonds. The O–H bonds in alcohols, for example, form particularly distinct peaks, although the spectrum can be complicated by the vibrations of nearby bonds interfering with each other. Just like other spectra, IR forms a molecular fingerprint that, with the right experience, can be read to determine a chemical compound's identity.

These molecular identification techniques are not just employed by chemists who have got their beakers mixed up. They can be used to monitor chemical reactions and identify large biomolecules with great enough accuracy to spot a change in a single amino acid in a large protein sequence. Mass spectrometry is widely used in drug discovery, drug testing, screening samples taken from newborn babies for certain diseases (see Newborn testing, page 85) and tracking down contaminants in food products.

Spectrum scandal

In chemistry, convincing proof of a reaction having taken place might hinge on an NMR spectrum. That proof might determine whether or not your paper gets published. With such high stakes, there are those who might be tempted to tweak the evidence to fit their argument. In 2005, Bengu Sezen, a chemist at Columbia University, USA, had several of her papers retracted after it was revealed she had cut and pasted the peaks of NMR spectra to get the results she wanted.

The condensed idea
Molecular fingerprints

22 Crystallography

Anything that involves firing X-rays at stuff automatically tends to sounds like science fiction – especially when you're using a multimillion pound piece of equipment to do it. Crystallography is very much within the realms of science fact, but that doesn't make it any less awesome.

Not far south of Oxford, England, surrounded by green fields, is a big shiny silver building. From the nearby road, it may look like a sports stadium, but if you ever happen to pass it, don't be fooled. Inside, scientists are accelerating electrons to unimaginable speeds to generate beams of light ten billion times brighter than the Sun. The building houses the Diamond Light Source, the most expensive scientific facility ever built in the UK.

A bit like the Large Hadron Collider, Diamond is a particle accelerator, except here the particles aren't crashed together, they're focused onto crystals a few thousandths of a millimetre across. Using Diamond's superbright light, scientists are able to peer into the heart of individual molecules and reveal how all the atoms are connected to each other.

X-RAY VISION

Diamond produces extremely powerful X-rays. Discovered by Wilhelm Röntgen in 1895, the X-ray is the basis for two centuries of pioneering work in understanding the structures of important biological molecules, as well

TIMELINE

1895	1913	1937	1946
Discovery of X-rays by Wilhelm Röntgen	William Bragg and son use X-rays to map atoms in a crystal	Hodgkin solves structure of cholesterol	Hodgkin solves structure of penicillin

as drugs, and even state-of-the-art materials being developed for solar panels, buildings and water purification. The theory is straightforward – the patterns formed when X-rays are scattered by a substance tell you how the atoms in the molecules are arranged in three dimensions. The scatter pattern is interpreted from a series of dots showing where the X-rays hit a detector.

Dorothy Crowfoot Hodgkin (1910–94)

Hodgkin is remembered as one of the outstanding scientists of the 20th century. She was also a lecturer, well-liked supervisor in her lab - where one of her students was the future British Prime Minister, Margaret Thatcher - Chancellor at the University of Bristol for many years and a supporter of humanitarian causes. Her face has appeared on designs for two British stamps.

Practically, however, it's anything but simple. The technique, called X-ray crystallography, depends on having perfect crystals – neatly ordered arrays of molecules. Not all molecules form perfect crystals easily. Ice and salt do it, but big, complex molecules like proteins have to be encouraged.

Just working out how to grow perfect crystals can take years and even decades. Such was the case when the Israeli chemist Ada Yonath decided to try to make ribosome crystals. The ribosome is the protein-making machine that produces proteins in cells. It's present in all living things, including microbes, meaning that knowing its structure could be valuable in fighting any number of dangerous diseases. The trouble is that ribosomes themselves are made up of lots of different proteins and other molecules, amounting to hundreds of thousands of atoms in total and a remarkably complex structure.

CRYSTAL METHODS

Starting in the late 1970s, Yonath tried for more than ten years to crystallize the ribosomes of various bacteria so she could bombard them with X-rays. When she finally had good enough crystals, the patterns produced by the rays weren't easy to interpret and the resolution on the images was quite low.

1956	1964	1969	2009
Hodgkin solves structure of vitamin B12	Hodgkin receives Nobel Prize for crystal structures of biological molecules	Hodgkin solves structure of insulin	Nobel Prize awarded for crystal structure of the ribosome

X-ray detecting

Today, scientists can glean structural information from crystals a fraction of the size of those that Dorothy Hodgkin was working with in the 1940s. That's because it's now possible to generate much more powerful X-rays. The X-rays are generated by high-speed electrons whizzing around in a particle accelerator. These electrons produce pulses of electromagnetic radiation that we refer to as X-rays. This is a type of electromagnetic radiation similar to visible light, but with a much shorter wavelength. Visible light can't be used to study atomic-level structure because its wavelength is too long – each wave is longer than an atom and therefore won't be scattered. During the experiment, crystals are mounted on the equivalent of a pinhead and kept cool while being subjected to the X-rays. The scattering of the X-rays is known as diffraction and the pattern they produce on the detector is called the diffraction pattern.

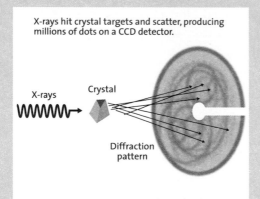

X-rays hit crystal targets and scatter, producing millions of dots on a CCD detector.

X-rays

Crystal

Diffraction pattern

It wasn't until 2000, after three decades and collaborating with other scientists with whom she would eventually share a Nobel Prize, that her images were finally sharp enough to reveal the atomic-level structure of the ribosome. It was a triumph nevertheless. When she started out, no one had believed it could be done. Recently, drug companies have been using the structures provided by Yonath and her colleagues to try to design new pharmaceuticals that could defeat drug-resistant bacteria.

Ada Yonath was not, however, the first woman to devote an entire career to crystallography. In fact, the entire field of X-ray crystallography was pioneered from the 1930s onward by Dorothy Crowfoot Hodgkin, who worked out the crystal structures of many important biological molecules, including cholesterol, penicillin, vitamin B12 and – after winning her own Nobel Prize – insulin. Despite being crippled by the pain of rheumatoid arthritis from age 24, she worked tirelessly to disprove her doubters. She studied penicillin during the Second World War, at a time when the technique was still new and viewed with suspicion by other researchers. At least one of her fellow Oxford University chemists is known to have scoffed at the structure she proposed – a structure that it would later be proved was correct. That structure she solved in just three years, while insulin took her more than thirty.

GOING DIGITAL

In Hodgkin's day, everything was done using photographic film – the X-rays hit the crystal and scattered onto a photographic plate that she would place behind. The spots on the film formed the pattern that she hoped would reveal the atomic-level structure. Today, X-ray crystallography is done using digital detectors, not to mention really powerful particle accelerators like the Diamond Light Source and computers capable of dealing with all the data and difficult calculations required to solve the structures. It was Hodgkin who campaigned for computers at Oxford, after she used those at Manchester University to help her solve the structure of vitamin B12. But until then, she had to use her formidable brainpower to compute the complex mathematics.

It seems X-ray crystallography and its supporters have come up trumps. Some scientists may once have doubted its use, but since the 1960s, crystallographic techniques have solved the structures of more than 90,000 proteins and other biological molecules (see page 152). X-ray crystallography is the go-to technique for studying atomic-level structure. Even though it is now well established, there are still inherent problems to overcome. Growing perfect crystals is never easy, so scientists have been working on ways of studying less-than-perfect crystals. And 60 years after Hodgkin started her lengthy studies on insulin, NASA scientists got a clearer look at it by growing it in space – far superior crystals can be grown in the microgravity environment of the International Space Station.

> **IF THAT'S THE FORMULA OF PENICILLIN, I'LL GIVE UP CHEMISTRY AND GROW MUSHROOMS.**
>
> Chemist John Cornforth, about Hodgkin's (correct) formula

The condensed idea
Revealing the structures of individual molecules

23 Electrolysis

At the turn of the 19th century, the battery was invented and chemists began experimenting with electricity. Soon, they were using a new technique called electrolysis to break substances apart and discover new elements. Electrolysis also became a source of chemicals, such as chlorine.

In 1875, an American doctor invented a method for destroying hair cells so he could zap his patients' ingrown eyelashes. He called his technique 'electrolysis', and it is still used to this day to remove unwanted body hair. However, this depilation method has very little to do with an equally pioneering electrolysis technique that was also employed in 1875 in the discovery of the silvery metal element gallium. Except for one thing – and the clue is in the name – both techniques require electricity.

By 1875, this second type of electrolysis had been around for half a century or more, and had already revolutionized 19th-century chemistry. It behoves us to never confuse this technique of experimental chemistry with the system for permanently removing leg hair. Electrolysis has also had a major impact in the realm of public health, eventually becoming the method used to extract chlorine from brine (chlorine is the disinfectant used in swimming pools and to keep our drinking water supply disease-free). At that time, though, it was probably better known as the method that renowned scientist and lecturer at the Royal Institution, Humphry Davy (see page 44), had used to separate a whole series of common elements, including sodium, calcium and magnesium, from their compounds.

TIMELINE

1800	LATE 1800s	1892
First description of battery by Alessandro Volta	Nicholson and Carlisle invent electrolysis	Electrolysis used industrially to produce chlorine from brine

Silver and gold plating

In silver or gold plating, electrolysis is used to form a thin layer of a more expensive metal on top of a more economical one. The metal object acts as one of the electrodes in what is called the electrolytic 'cell'. You could silver-plate a spoon by attaching it to a wire and battery, and dunking it in some silver cyanide dissolved in water. The spoon becomes the negative electrode and the positively charged silver ions in the water are attracted to it. To maintain the supply of silver ions, a piece of silver is used as the positive electrode. In effect, the silver is transferred from one electrode to another. In the same way, gold can be attached to a wire and used to coat jewellery or a smartphone case, for example. The solution that the electrodes are dunked in is called the electrolyte.

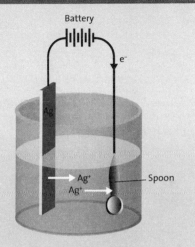

SPLITTING WATER

Although Davy was the most famous electrolysis experimentalist, credit for its invention goes to a little-known chemist by the name of William Nicholson and his friend, surgeon Anthony Carlisle, in 1800. They were fascinated by some experiments that battery pioneer Alessandro Volta had completed earlier that year and were trying to repeat them. At this point, Volta's 'battery' was just a pile of metal discs and wet rags with wires attached. Intrigued to find that bubbles of hydrogen appeared when a battery wire touched a drop of water, they took the wires and placed them in either end of a tube of water. The result was oxygen bubbles at one end and hydrogen at the other. They'd used electricity to break the bonds between molecules of water, splitting them into their component parts.

1854

John Snow shows that water can spread disease

1908

Chlorine used in water supply for first time

Nicholson being an accomplished lecturer, writer and translator, who had already set up his own popular science journal, had no doubt about where he was going to publish their results. The *Journal of Natural Philosophy, Chemistry and the Arts*, known affectionately as Nicholson's Journal, soon featured an article heralding the dawn of a new era of electrochemistry..

ELECTROCHEMISTRY

Volta's pile was adopted and adapted – eventually, it became something resembling a modern battery – and soon scientists were using electrolysis for all sorts of interesting chemistry. Davy isolated calcium, potassium, magnesium and other elements, while his Swedish rival Jöns Jakob Berzelius worked on splitting various salts dissolved in water. In chemistry, a salt means a compound made up of ions whose charges cancel each other out. In table salt – sodium chloride – the sodium ions are positively charged and the chloride ions are negatively charged. Sodium can also form a bright yellow salt with chromate (CrO_4^-) ions. While it is far more exciting-looking than table salt, sodium chromate is also toxic and inedible.

> ❝ THE GRAND QUESTION RESPECTING THE DECOMPOSITION OF WATER ... DERIVES POWERFUL CONFIRMATION FROM THE EXPERIMENTS FIRST PERFORMED BY MR NICHOLSON AND MR CARLISLE ... ❞
>
> John Bostock
> in 'Nicholson's Journal'

This brings us neatly to our modern-day understanding of how electrolysis actually works, because it's all about the ions (see Ions, page 19). When a salt is dissolved in water it disintegrates into its positive and negative ions. In electrolysis, those positive and negative ions are attracted to the electrodes with the opposite charges. Electrons are entering the circuit at the negative electrode, so positive silver ions (see Silver and gold plating, page 93), for instance, pick up electrons to form a coating of neutral silver atoms. Meanwhile, the negative ions attracted to the other electrode do the opposite – they lose their extra electrons to become neutral.

Certain salts, such as standard table salt, contains sodium ions, which although positively charged like silver ions, are more reactive. So when sodium ions are split from chlorine they team up immediately with hydroxide ions (OH–) in the water electrolyte and form sodium hydroxide.

Instead of the negative electrode attracting sodium ions, it attracts hydrogen ions, which collect electrons and bubble off as hydrogen gas.

CLEAN REVOLUTION

The same set-up forms the basis for an entire industry of chlorine production by electrolysis. Basically, run an electric current in seawater and you can collect chlorine. The by-product, sodium hydroxide, also known as caustic soda, can be combined with oil to make soap.

At the same time as electrochemistry was progressing in the 19th century, scientists were becoming increasingly aware of the problems of waterborne disease. Until about the middle of the century, cholera was thought to be contracted by breathing the miasma of 'bad air'. During an outbreak of cholera in London in 1854, however, John Snow showed that people were being infected by dirty water from a pump in Soho by plotting cases on a map, and thus asserted himself as one of the first epidemiologists.

Electricity

The 'voltaic pile' invented by Alessandro Volta provided the first steady supply of electricity. Before that, foil-lined Leyden jars were used to trap and store electricity discharged as a spark from a hand-cranked static electricity generator. The jars were filled with water or even beer to store the electricity – until scientists realized it was actually the foil, not the liquid, that was storing the charge.

Within a few decades, chlorine, produced by electrolysis, was being used as a disinfectant to protect people from microbes in their drinking water. It was used for the first time to treat the water supply of Jersey City, New Jersey, in the USA. Chlorine is also used in bleach, and many drugs and insecticides. Today, the hydrogen bubbles formed in the electrolysis of salt-water are sometimes collected and used in fuel cells to generate yet more electricity.

The condensed idea
Electricity breaks down chemical compounds

24 Microfabrication

You might have tens or even hundreds of computer chips in your home and while each of these is an incredible feat of engineering, it is also the result of some important chemical advances. It was a chemist who etched the first patterns into silicon wafers and though the chips might be much smaller today than they were 50 years ago, the chemistry of silicon remains the same.

Few technologies have had as profound an impact on human society and culture as the silicon chip. Our lives are ruled by computers, smart phones and a myriad other electronic devices driven by integrated circuits – chips or microchips. The miniaturization of electronic circuits and devices has quite literally put computing power in all of our pockets, shaping the way we experience the world today.

Yet one of the key chemical advances that led to the development of the silicon chip is sometimes glossed over. Historical accounts never fail to credit Jack Kilby of Texas Instruments, who later won the Nobel Prize in Physics, as the inventor of the integrated circuit and refer repeatedly to Bell Laboratories – Bell Labs – where the first transistors were made, but Bell Labs' chemist Carl Frosch and his technician, Lincoln ('Link') Derick, often only get the briefest of mentions.

FRESHMAN FROSCH

Perhaps this is because not much is known about Frosch. Very little is written about his early career or personal life. He was recognized as a scientific

TIMELINE

1948	1954	1957
First transistor unveiled by Bell Labs	Carl Frosch and Lincoln Derick grow a silicon dioxide layer on a silicon wafer	Bell Labs use a 'photoresist' to transfer a pattern onto a silicon surface

talent at a young age – a grainy black and white image of a brooding, 21-year-old Frosch appears in the 2nd March 1929 edition of New York's *Schenectady Gazette*, next to an advert for 'Extra fancy Mohican Sifted Peas'. The accompanying article announces his election to the Sigma XI honorary scientific fraternity, the 'highest honour' that can be bestowed upon a science student – but then things go quiet for more than a decade.

By 1943, Frosch was working for Bell Labs at its Murray Hill Chemical Laboratories. A colleague, Allen Bortrum, remembers him as a modest man, although he must also have had a competitive streak because he's pictured in the June edition of the *Bell Laboratories Record* receiving a trophy for the highest score in the Murray Hill Bowling League. Five years later, Bell Labs unveiled the first transistor, made of germanium. Tinier versions of these miniature electronic switches would eventually be crammed onto modern

Making chips

One of the first simple patterns that Frosch etched into his wafers was 'THE END'. In basic terms, the process of making an integral circuit or computer chip is a bit like printing combined with developing a photograph. In fact, it was printing technology previously used to make patterns on printed circuits boards that was adapted to transfer designs onto silicon wafers. Now, it is possible to etch very complex designs and use multiple masks on the same silicon wafer.

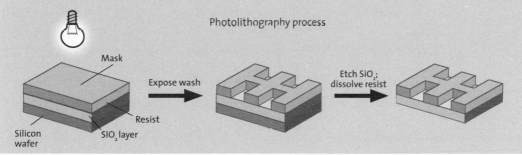

Photolithography process

1958
Jack Kilby at Texas Instruments invents the integrated circuit

1965
Moore's Law first stated in *Electronics* magazine

1965
Number of electrical components on a computer chip reaches one billion

Doping

Silicon has four electrons in its outer shell. In a silicon crystal, each silicon atom shares these four electrons with four other silicon atoms – a total of four shared pairs per atom. Phosphorus has five electrons in its outer shell, so when it is added as a dopant it provides a 'free' electron that wanders around the silicon crystal and can carry a charge. This type of doping creates so-called 'n-type' silicon – the charge carriers are electrons (negative charge). The other type is 'p-doping' – p equals positive charge. Here, the charge is carried by the absence of electrons. This might sound like an odd concept, but consider that boron – an n-type dopant – has one less electron than silicon in its outer shell. This means there is a gap, or electron 'hole', in the crystal structure where an electron was supposed to be. Positively charged holes also carry charge by accepting electrons.

computer chips in their millions and billions, but these would be made of silicon. It was Frosch and Derick, a former fighter pilot, who made the discovery that earned Silicon Valley its name.

FLASH IDEAS

By the 1950s, transistors were being made by a method called the diffusion process, where dopants – chemicals that alter the electrical properties of a substance – were introduced by diffusion in gases into extremely thin wafers of germanium or silicon at very high temperatures. There was still no such thing as an integrated circuit at this point. At Bell Labs, Frosch and Derick were focusing on improving the diffusion method. They were already working with silicon, since germanium was prone to defects, but they didn't have the best equipment and Frosch was constantly cremating the silicon wafers.

Their experiments involved placing a wafer in a furnace and directing a stream of hydrogen gas containing a dopant at it. One day, Derick returned to the lab to find that the hydrogen stream had set alight to their wafers. On inspecting the wafers, however, he was surprised to find that they were bright and shiny – oxygen had leaked in, causing the hydrogen to burn to produce steam. The steam had reacted with the silicon to produce a glassy top layer of silicon dioxide. This silicon dioxide layer is not central to photolithography – the method still used to make silicon chips.

RINSE AND REPEAT

In photolithography, the pattern for the integrated circuit is etched into the silicon dioxide layer. It is covered by what is called a photoresist – a photosensitive layer – and on top of this, a mask, which contains a repeated pattern so that many chips can be made at once. Beneath the mask, the exposed areas of the photoresist react to light and can be washed away to reveal the transferred pattern. This pattern is then etched into the shiny silicon dioxide layer below.

What Frosch and Derick realized was that they could use the silicon dioxide layer to protect a wafer from damage in the high temperature diffusion process and to define the areas they wanted to dope. Boron and phosphorus dopants (see Doping, opposite) can't get through the silicon dioxide layer, but by etching windows into the layer, it was possible to carry out diffusion of dopants in very specific spots. In 1957, Frosch and Derick published a paper in the *Journal of the Electrochemical Society* detailing their discoveries, noting the potential for making 'precise surface patterns'.

Semiconductor firms quickly latched onto the idea. They were trying to make multiple transistors from single wafers. Then, just a year later, Kilby invented the integrated circuit – a device where all the components were made simultaneously from one slice of semiconductor material. This 'chip' was actually germanium-based; however, a germanium dioxide layer does not act as a barrier, so silicon eventually caught on. Today, extremely complex patterns are designed on computers and transferred on to silicon wafers using the oxide masking method. In 1965, Intel founder Gordon Moore predicted that the number of components on a computer chip would double every year, later revising his prediction to every two. Thanks to advances in photolithography, we've managed to keep up the pace, with the one billion mark being surpassed in 2005.

> **SILICON ITSELF IS, OF COURSE, THE CRITICAL INGREDIENT, FOLLOWED BY ITS UNIQUE NATURAL OXIDE, WITHOUT WHICH LITTLE OF TODAY'S THRIVING SEMICONDUCTOR INDUSTRY WOULD EVER HAVE BEGUN TO EXIST.**
>
> Nick Holonyak, Jr., inventor of the LED

The condensed idea
Silicon chemistry
in every smartphone

25 Self-assembly

Molecules are too small to see through ordinary microscopes, so scientists are limited in their ability to manipulate them with ordinary tools. What they can do instead is redesign the molecules so that they organize themselves. Self-assembling structures could be used to create miniature devices and machines straight from the pages of science-fiction books.

f you had to make your own spoon, how would you do it? What would be your first instinct? Would you try to find a lump of metal, or perhaps a tree branch, and bash or carve it into the right shape? That, perhaps, would be the most obvious way, but it wouldn't be the only way. An alternative method – although it might initially seem more tedious – would be to collect hundreds of tiny scraps of metal, or woodchips, and stick them together in the shape of a spoon.

The first way of doing it is what chemists call a 'top-down' approach. You take a bulk material and carve something out of it in the shape and size that you want. The second way of doing it is the opposite – 'bottom-up'. Instead of paring down the bulk material, you work up from smaller pieces. True, the second way sounds like an awful lot of hard work; however, imagine if, rather than having to stick all the pieces together yourself, they did it themselves. This would make things a lot easier.

WORKS LIKE MAGIC

This is basically what happens in molecular self-assembly, except on a much smaller scale. In nature, nothing is made from the top down. Wood,

TIMELINE

1955	1983	1991
Tobacco mosaic virus self-assembled in a test tube	First self-assembling monolayer made on a gold surface with alkyl thiolate molecules	Nadrian Seeman's group self-assemble a DNA cube

bone, spider's silk – all these materials are assembled, molecule by molecule, and they form spontaneously. When the outer membrane of a cell forms, for example, the lipid particles that make up the membrane organize themselves into a layer that becomes an envelope for the cell.

If we could somehow devise ways of making things that self-assembled, like in nature, from the bottom up, it would be nothing short of magic – like a sequence from the *Harry Potter* films where, with a spell and a quick flick of a wand, everything flies into place. We could build computer components, molecule by molecule – chips so small you could fit the computing power of NASA in your mobile phone (well, almost). We could make medical machines capable of going into our bodies to scrape our arteries clean, diagnose cancer or deliver a drug right to the site of an infection.

All this might sound far-fetched, but some of it is already happening. In labs around the world, scientists are coming up with self-assembly schemes where the particles get together of their own accord. Either they are guided intro place by moulds or patterns made by more traditional 'top-down' techniques, or the structures that they are intended to form are actually encoded into the very particles themselves. Such schemes are often the design of those working in the field of nanotechnology (see page 180). Self-assembling molecules can be used to create extremely thin layers of specialized materials and extremely tiny devices. The materials and structures that nanotechnologists are making are on a tiny scale – in the realm of one

Self-assembled monolayers

Self-assembling monolayers are one-molecule-thick layers that form in a well-ordered manner on a surface. The effect was first used in the 1980s to assemble alkylsilane and then alkanethiol molecules onto a surface. The sulfur in an alkanethiol molecule has a strong affinity for gold, so it will stick to a gold surface. By tailoring the rest of the molecule, it's possible to create thin films with diverse chemistries. For instance, antibodies or DNA can be attached making the films useful for medical diagnostics

2006

Paul Rothemund reports first folding of DNA like origami

2013

UK researchers develop MRSA test based on self-assembled monolayer for detecting bacterial DNA

Self-assembly in liquid crystals

The molecules in most modern TV screens are in a liquid crystal state (see page 24), in which there is a certain degree of regular arrangement combined with liquid-like flow. The molecules naturally assemble in a certain way but applying an electric field changes their arrangement, to control what we see on a display. Scientists have identified many natural materials that behave like liquid crystals and that self-assemble. For example, materials that make up the tough cuticles of certain insects and crustaceans are considered to form by liquid-crystalline self-assembly. New ways of manipulating the arrangements of such substances may be interesting for creating novel materials. In one 2012 study, Canadian scientists showed that using cellulose crystals produced from spruce wood they could form an iridescent film that was capable of encrypting secure information under different lighting conditions. Another study used liquid-crystalline cellulose to make a tiny humidity-driven 'steam engine'. Moisture changed the arrangement of crystals in a belt of cellulose film, causing strain that pulled on the wheel to make it rotate.

Moisture-driven cellulose motor

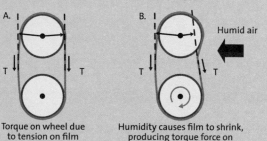

A. Torque on wheel due to tension on film is equal

B. Humidity causes film to shrink, producing torque force on wheel, turning it clockwise.

Humid air

millionth of a millimetre – so it makes more sense to build them up, molecule by molecule, than to use materials and tools that are giant by comparison.

FOLDS LIKE ORIGAMI

Clearly, you wouldn't want to make a normal-sized spoon this way, but if you wanted to make a nano-sized spoon, it would definitely be the way to go. Scientists at Harvard University in the USA have even gone one better. In 2010, they made what lead chemist William Shih referred to as, 'little Swiss Army knives' from self-assembling molecules. They borrowed from nature itself by using strands of DNA (see page 140) that folded into three-dimensional structures. Although they called them Swiss Army knives, these structures were more like tiny tent frames, with struts and staples providing incredible strength and rigidity. The scientists were able to make exactly the structures they wanted by designing the DNA code so that the molecules would only fold up in a certain way.

Far from being the first example of nanoscale engineering using DNA, the team had built on the work of others practising the art of what is widely called 'DNA origami'. While there may be no obvious use for tiny tent frames, the analogy with origami provides some hint of the level of possibility. Just as the same piece of paper can be folded into a beautiful bird or a stinging scorpion, DNA has the versatility to adopt any shape or structure – as long as its designer is capable of encoding that design into the DNA sequence.

Shih and his team are bioengineers. They work with biological materials and they try to solve biological problems. So they plan on developing their wire-frame structures to use in the human body, taking advantage of their biocompatibility. For example, their strength and rigidity may be useful in regenerative medicine, in the repair or replacement of damaged tissues and organs with lab-made tissue scaffolds. Meanwhile, scientists from electronics backgrounds are using other materials to develop self-assembly schemes for tiny sensors and low-cost electronics.

> **IT'S THE DIFFERENCE BETWEEN BUILDING NANOSCALE STRUCTURES, MOLECULE BY MOLECULE, USING THE EQUIVALENT OF NANO-CHOPSTICKS, AND LETTING MOLECULES DO WHAT THEY DO BEST, SELF-ASSEMBLE THEMSELVES ...**
>
> John Pelesko

THE ART IN THE SCIENCE

As a method, self-assembly may work like magic, but it takes a very skilled scientist to get it to work. Strictly speaking, the self-assembly part is hardly a method at all. It's just something that happens after all the hard work has already been done. The real art is in designing molecules, materials and devices so that they will self-assemble. Scientists aren't just making spoons – they're designing materials that will make the spoons themselves.

The condensed idea
Molecules that
organize themselves

26 Lab on a chip

Lab-on-a-chip technology has the potential to change the way medicine works by offering on-the-spot tests, for everything from food poisoning to Ebola, which can be carried out without any specialist knowledge. It's already possible to run hundreds of experiments at once on a tiny chip, small enough to fit in your pocket.

You go to the doctor with some mystery stomach bug and hope in vain that they won't say those dreaded words, 'I'm going to need a stool sample.' Yes, at some point in our lives, most of us will probably have to collect a plastic vial of our own waste and make an awkward delivery to the clinic. Thankfully, once you've made the delivery, it will be whisked straight off to the lab and you don't ever have to see it again. In the not-too-distant future, though, your doctor may be able to analyse your sample right in front of your eyes and give you the results within 15 minutes.

In 2006, US researchers working on a National Institutes of Health-funded project reported that they were developing a 'disposable enterics card' that could distinguish between bugs like *E. coli* and *Salmonella* by running a series of parallel tests on a stool sample – all on a single microchip. Their device would use antibodies to detect molecules on the surface of a microbe and then extract and analyse its DNA.

It sounds incredibly clever, if slightly revolting. But the enterics card is not a one-off. So-called 'point-of-care tests' may be the next big thing in healthcare and many of them are based on 'lab-on-a-chip' technology. Devices already

TIMELINE

1992	**1995**	**1996**
Microchip technology applied to make a microdevice for separating molecules in tiny glass capillaries	First use of a microdevice to sequence DNA	*Salmonella* DNA detected on a chip

exist for diagnosing heart attacks and monitoring T-cell counts in HIV patients. Cheap diagnostic chips could one day play a crucial role in monitoring the spread of epidemics. The big advantage of using one of these chips is that it doesn't require any specialist knowledge – it's an automated experiment that fits into the palm of your hand. All a doctor has to do is add a small amount of your sample and insert the card into a card reader.

MICROCHIP MEETS DNA

The lab-on-a-chip concept emerged when scientists began to realize they could hijack conventional microchip fabrication technology (see page 96) to create miniaturized versions of standard laboratory experiments. In 1992, Swiss researchers showed they could carry out a common separation technique called capillary electrophoresis (see page 82) on a chip device. By 1994, chemist Adam Woolley's team at the University of California, Berkeley, USA, was already separating DNA in tiny channels on a glass chip and, soon after, they used chips to carry out DNA sequencing. Today, DNA sequencing on glass and polymer chips has become perhaps the most important application of lab-on-a-chip technology, with chips capable of sequencing hundreds of samples in parallel and producing results in minutes.

Sequencing on a chip is no mean feat. It usually involves a technique called the polymerase chain reaction (PCR) that molecular biology has used for many years. The reaction depends on repeatedly heating and cooling DNA. To achieve this on a chip, samples in the channels have to be heated or forced through successive reaction chambers – each with a total volume below one-thousandth of a millilitre – at different temperatures. One major area of lab-on-a-chip technology is known as microfluidics. Because of the tiny volumes of liquid involved, most diagnostic chip devices are based on microfluidics.

Detective work

Fast, on-chip analysis of chemicals could also be useful for exposing foul play, for example, in testing drugs cheats or identifying rogue ingredients in cases of food adulteration. A lab-on-a-chip device could run a test for many different illicit drugs or, in sport, banned substances, and provide an answer in minutes.

1997
DNA sequencing in parallel lanes on a microchip

2012
Prediction of lab-on-a-smartphone technology for medical monitoring

2014
Concept for 'Internet of Life' announced

The Internet of Life

You may have heard of the 'Internet of Things', a concept that draws on the idea that we are living in a world of increasingly smart devices that could all be connected by a single network. Smartphones, refrigerators, TVs and even microchipped dogs, could all be integrated into the network - via their microchips. Now researchers at QuantuMDx, a company based in Newcastle-upon-Tyne, England, are planning an 'Internet of Life', which would integrate data produced by lab-on-a-chip devices all over the world. They suggest that DNA-sequencing data collected with chip devices could be geostamped, meaning it could be mapped to a specific geographical location. This would give epidemiologists access to unprecedented levels of detail for tracking disease in real time. They could monitor malaria, follow the evolution of the flu virus, help to predict Ebola outbreaks, identify new strains of drug-resistant tuberculosis and, hopefully, use all this information to help stop the spread.

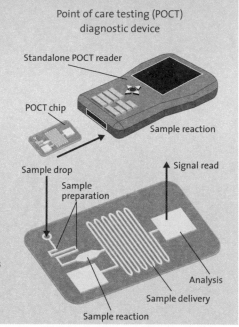

Point of care testing (POCT) diagnostic device

Standalone POCT reader

POCT chip

Sample reaction

Sample drop

Sample preparation

Signal read

Analysis

Sample delivery

Sample reaction

There are, however, many other uses for these chip-based technologies. From a chemist's perspective, the channels and chambers in a chip provide a way of carrying out reactions and analyses in a controlled and repeatable way, using sample sizes too small for human hands to deal with. Biologists can capture single cells within individual reaction chambers and test the effects of different chemicals or biological signalling molecules simultaneously. Drug developers could use them to mix tiny quantities of different drugs to aid in testing of their combined effects. In all of these areas, working with such small amounts helps keep waste and costs to a minimum.

Chips could also be useful for formulating and delivering drugs, for example, creating micro- or nano-sized capsules, or measuring out and drip-feeding miniscule doses to reduce the side effects associated with sudden surges in medication levels. Some experts envisage patients carrying portable drug delivery chips. These could even be attached via 'microneedles' to the target tissues, such as at the site of a tumour.

NETWORKED DISEASE DATA

Diagnostics and personal health monitoring, however, are some of the most exciting areas for those working in lab-on-a-chip technology. The molecules most commonly tested in lab-on-a-chip devices are proteins, nucleic acids like DNA and molecules involved in metabolism. Chips have a very obvious application for diabetics, who must constantly monitor their blood-sugar levels (see Sugar sensing, page 136). There are other so-called 'biomarker' proteins that can indicate many conditions, such as brain damage or tell a midwife if a woman is going into labour. Very often, diagnostic chips make use of antibodies because they are good at recognizing specific molecules – our own as well as those belonging to infectious organisms.

> **[THERE] IS A LOT OF TECHNOLOGY TODAY WHERE IT ACTUALLY BYPASSES TRADITIONAL PHYSICIAN INVOLVEMENT ... WE'RE TALKING ABOUT LAB-ON-A-CHIP, ON A PHONE ...**
>
> Eric Topol, Director of the Scripps Translational Science Institute on the *Clinical Chemistry* podcast

Chip diagnosis could make a much more important impact in areas of the world where resources are scarce and facilities for professional lab analysis of samples might not be available. One UK-based company wants to feed results from its diagnostic device to a networked database, creating an 'Internet of Life' (see The Internet of Life, opposite) that could monitor outbreaks of deadly diseases like Ebola. So while it might be a few years before you find yourself at the doctor, getting an on-the-spot stool analysis, lab-on-a-chip devices could one day lead to a revolution in the way we deal with disease. And as we'll see elsewhere, computer power has many other uses in chemistry.

The condensed idea
Chemistry experiments in miniature

27 Computational chemistry

A birdspotter and biologist at heart, Martin Karplus might have seemed like an unlikely candidate for the father of computational chemistry. However, he believed theoretical chemistry could provide a basis for understanding life itself, and so it proved – he just had to get past a five-ton computer first.

Martin Karplus, the father of computational chemistry, was an Austrian Jew whose family left Austria for the USA in 1938 – as Austria was joining with Nazi Germany. At school in the USA, Karplus was recognized as a bright student. Outside school, his interest in science grew alongside his passion for nature. He was a young 'birder', recording sightings for the Audubon Society's annual bird migration census. Aged 14, he was nearly arrested on suspicion of being a German spy signalling to submarines – he was out in a storm with a pair of binoculars, looking for little auks.

Before attending college, Karplus was invited to take part in some research on bird navigation in Alaska and became convinced a research career was for him. However, instead of enrolling on a biology course, he signed up for Harvard's Chemistry and Physics program, reasoning that these subjects would be critical to eventually understanding biology and life itself. As a PhD student at Caltech, he started a project on proteins, but his supervisor left and he was adopted by Linus Pauling – who would soon win the Nobel Prize for Chemistry for his work on the nature of the chemical bond. Karplus

Computers in drug research

To research whether a newly designed drug does what it is supposed to do, it needs to be tested. But with hundreds or thousands of different potential drugs and a limited workforce and limited money, it's almost impossible to test all of them in real cells, animals or people. This is where computational chemistry comes in. Using molecular simulations, it's possible to work out how drug molecules might interact with the molecules in the body that they target and therefore identify the drugs that make the best candidates for tackling a particular disease. These theoretical calculations can be thought of as experiments *in silico* - in silicon, or computers. Of course, there may be problems with

the drugs that the simulations don't pick up, but this is why the combination of computational (theoretical) and experimental chemistry is so strong.

A computer prediction of protein structure compared to crystallographic evidence

studied hydrogen bonding (see page 20) and was forced to write up his thesis in just three weeks when Pauling suddenly announced he was leaving on an extended trip.

After a spell with a theoretical chemistry group at Oxford University, Karplus settled for five years in a position at the University of Illinois, working on nuclear magnetic resonance (NMR) (see page 84). He was using NMR to study the bond angles of hydrogen atoms in the ethanol (CH_3CH_2OH) molecule, when he realized that doing all his calculations on a desk calculator was going to be very tedious – so he wrote a computer program to do the work for him.

1977

First molecular dynamics simulation of a large biological molecule – bovine pancreatic trypsin inhibitor (BPTI)

2013

Martin Karplus, Michael Levitt and Arieh Warshel awarded Nobel Prize for computational chemistry

> **THEORETICAL CHEMISTS TEND TO USE THE WORD "PREDICTION" RATHER LOOSELY TO REFER TO ANY CALCULATION THAT AGREES WITH EXPERIMENT, EVEN WHEN THE LATTER WAS DONE BEFORE THE FORMER.**
>
> Martin Karplus

THE FIVE-TON COMPUTER

At that time, in 1958, the University of Illinois was the proud owner of a five-ton digital computer called ILLIAC, which had a full 64 KB of memory – not enough to hold a single digital photo taken on your mobile phone, but enough for Karplus' program – and was programmed by punch cards. Soon after finishing the calculations, he attended a talk by one of the organic chemists at Illinois who seemed to have confirmed his results experimentally.

Filled with confidence that his calculations could be useful in determining chemical structures, Karplus published a paper that included what has become known as the Karplus equation. The equation was used by chemists to interpret NMR results and determine molecular structures for organic molecules. His original formulation of the equation was refined and adapted, but it is still used in NMR spectroscopy today. The lecture that Karplus had attended was on sugars but his equation has been extended to other organic molecules, including proteins, as well as to inorganic molecules.

In 1960, Karplus moved to the IBM-funded Watson Scientific Laboratory, which had an IBM computer that was faster and had more memory than the ILLIAC. Realizing very quickly that an industry career wasn't for him, he returned to academia but with something that would help his research progress: access to the IBM 650. He continued working on problems that had intrigued him while at Illinois. Except now, he had the means to really attack them, using the IBM computer to help him probe chemical reactions at the molecular level.

BACK TO NATURE

Eventually, Karplus returned to Harvard and to his first love, biology. Here, he applied his now-considerable experience in theoretical chemistry to animal vision. Karplus and his team suggested that one of the C–C bonds in retinal – a form of vitamin A that detects light in the eye – twisted when exposed to light and that this movement was key to vision. Their theoretical calculations predicted the structure that the twisting action would produce. The same year, experimental results proved they were right.

Theoretical results from computational chemistry often go hand in hand with empirical evidence. The theory underpins the observations, just as the observations support the theory. Together, they make a much more compelling case than either alone. After Max Perutz produced crystal structures for haemoglobin – the oxygen-carrying molecule in blood – Karplus came up with a theoretical model to explain how the two interacted.

Uniting biology, chemistry... and physics

Not only did Martin Karplus have to learn chemistry to explain biology, he had to unite chemistry with physics to do it. The Nobel Prize (in Chemistry) that Karplus and colleagues were awarded in 2013 (see page 109) was given for harnessing both classical and quantum physics to develop the powerful models that would allow chemists to model really large molecules – like those found in biological systems.

DYNAMIC FIELD

Karplus went on to study how protein chains fold to form working protein molecules and to work with his graduate student, Bruce Gelin, on developing a program that would help calculate protein structures from a combination of amino acid sequences and X-ray crystallography (see page 88) data. The resulting CHARMM (Chemistry at HARvard Macromolecular Mechanics) initiative in molecular dynamics is still going strong.

Today, models and simulations are almost as important to the field of chemistry as they are to economics. Chemists are developing computer models that can simulate reactions and processes such as protein folding at the atomic level. These models can be applied to processes that would be nearly impossible to catch in action because they are happening on the scale of fractions of a second.

The condensed idea
Modelling molecules with computers

28 Carbon

Carbon is the chemical element blamed for destroying the environment. Yet it's also the basis for life on Earth – everything that ever lived was made of carbon-containing molecules. How has one little atom sneaked into every corner of the planet? And how can two compounds containing nothing but carbon look completely different?

If there's one element we hear about perhaps more than any other, it's carbon. Most of what we hear is bad, of course – carbon is clogging up the atmosphere and throwing Earth's climate into disarray. The constant focus on curbing carbon emissions means we think of carbon as a force to be tamed. So it's easy to forget that carbon itself is just a tight little ball of protons and neutrons surrounded by a cloud of six electrons. A simple chemical element resting above silicon on the Periodic Table. So aside from its environmental misdemeanours, what's so important about carbon that means it deserves special attention?

What we sometimes neglect is that carbon is the basis of everything living here on Earth – everything that creeps, crawls, flaps and flies. It's carbon that forms the chemical backbone for all biological molecules, from DNA to proteins, and from fats to the neurotransmitters flitting between the synapses in our brains. If you could take every atom in your body and count them out, more than one in six would be carbon. There would only be more oxygen atoms because most of your body is water.

TIMELINE

1754	1789	1895	1985
Joseph Black discovers carbon dioxide	Antoine-Laurent Lavoisier proposes the name carbon	Svante Arrhenius presents paper on effects of atmospheric carbon	Buckyballs created in the lab

ORGANIC AND INORGANIC

The extraordinary diversity of carbon-containing compounds is down to carbon's willingness to bond with itself – as well as other atoms – and to form rings, chains and other sophisticated structures. Nature, on its own, is capable of making millions of different complex carbon compounds. Many will probably disappear before we even discover them, as the plants, animals or bugs that make them go extinct. With the addition of human ingenuity, the possibilities for making new carbon compounds synthetically are virtually infinite.

All of these carbon compounds fall under the scope of what chemists refer to as organic chemistry. Their 'organic' label might fool you into thinking they ought to be limited to compounds made by nature and this is, in fact, how organic chemicals were initially classified. But today we recognize plastics as organic chemicals in the same way that we recognize proteins as organic chemicals, because they both contain carbon skeletons. Almost all carbon-containing compounds, with some notable exceptions, are organic, regardless of whether they're made in a beetroot, a bacterium or at the bench in a chemistry lab.

In general, anything that isn't organic is inorganic. Like organic chemistry, inorganic chemistry has its subdivisions, but it's a mark of the importance of carbon that chemistry is carved up in this way. One of the most obvious outcasts is the molecule clogging up our atmosphere – carbon dioxide. It doesn't really fit into any subdivision. Although it does contain carbon, it doesn't have what chemists would call 'functional groups'. Most organic compounds can be further subdivided based on which groups of atoms hang off their carbon skeletons. But, since carbon dioxide only has a couple of oxygen atoms, it is left it in a bit of an odd, 'in-betweeny' place.

2009

110 world leaders gather at Copenhagen climate talks to discuss action on climate change

2010

Nobel Prize in Physics awarded for getting graphene from graphite

There is a whole class of exceptions known as organometallics. These are carbon-containing compounds in which certain carbons are bonded to metals. Organometallic compounds are viewed as somewhere in between organic and inorganic, and they are more often the domain of inorganic chemists. These are not particularly obscure chemicals by any stretch of the imagination. Neither are they only made exclusively in chemistry labs. The haemoglobin molecules that carry the oxygen around in your blood harbour iron atoms and vitamin B12 contains a cobalt (see page 48). Like B12, organometallic compounds tend to make good catalysts.

> **THE SLIGHT PERCENTAGE OF [CARBON] IN THE ATMOSPHERE MAY, BY THE ADVANCES OF INDUSTRY, BE CHANGED TO A NOTICEABLE DEGREE IN THE COURSE OF A FEW CENTURIES.**
>
> Svante Arrhenius, 1904

CARBON-ONLY COMPOUNDS

Another odd carbon compound is diamond, which is all carbon and yet isn't considered organic. (It's sometimes best not to question the categorization systems of chemists.) There are several fascinating pure-carbon compounds with which it's worth becoming acquainted. As well as diamond, there's carbon fibre, carbon nanotubes, buckyballs, pencil lead (graphite) and a chicken-wire structured, atomically thin carbon compound called graphene that chemists are hoping will be the next big thing in electronics (see page 184).

What's weird is that if you think about diamond and pencil lead, they seem to bear no resemblance to each other (see Diamond vs. pencil lead, opposite). They are both wholly made up of carbon atoms, just arranged in different ways. As a result of their different atomic structures – the way the atoms are bonded together – they have completely different appearances and properties. Graphene, on the other hand, isn't so different, structurally, to graphite. It's actually possible to use sticky tape to pull the one-atom-thick sheets of carbon off a lump of pencil lead.

CARBON UNLEASHED

All this interesting and useful chemistry doesn't get carbon off the hook. Or rather, it doesn't get us off the hook. The fossil fuels that we burn to get energy are hydrocarbons, and when carbon-containing fuels such as petrol

Diamonds vs. pencil lead

In diamond, each carbon atom is bonded to four others, whereas in graphite, each carbon atom is only bonded to three others. While the bonds in diamond stretch out in different directions, in graphite they form a flat plane. This means that the structure of diamond is a rigid three-dimensional network while graphite is formed of a stack of loosely connected carbon layers. The layers in pencil lead are held together by weak attractions called van der Waals forces, but these are broken easily – just pressing pencil to paper is enough to release the uppermost layer. These molecular-level structural differences make diamond very hard and graphite, comparatively, very soft.

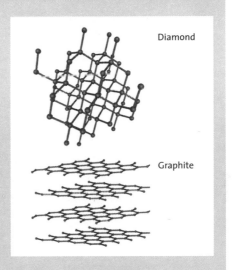

Diamond

Graphite

and coal, are burned, the combustion reaction produces carbon dioxide. This combustion reaction releases carbon that has been locked up under the ground for millions of years into the atmosphere, where it works to stop infrared radiation escaping into space – a process called the greenhouse effect, which is helping to drive global warming. Irrespective of the role carbon plays in our bodies, or in the lead of a pencil or in potential electronic devices of the future, the fact that we're unleashing billions of tonnes of the stuff every year remains a massive problem.

The condensed idea
One element, many faces

29 Water

You wouldn't think water had too many secrets – you can see right through it, for one thing – but water has hidden depths: if carbon compounds are the stuff of life, then water is the medium in which it thrives and survives. Yet despite decades of research into its structure, there's still no model that can tell us exactly how water behaves in every situation and why.

H_2O is perhaps the only chemical formula, besides CO_2, that the vast majority of people would be able to quote without taking a second to think about it. If there is one chemical that ought to be easy to understand then water is it. However, understanding the water molecules that burst from our taps, fill the ice trays in our freezers and keep our lakes and swimming pools wet has proved very far from simple. While we might think of water more as a backdrop for our holiday snaps than as a chemical, a chemical is exactly what it is, and a complicated one at that.

> **THE GREATEST MYSTERY IN SCIENCE IS UNDERSTANDING WHY, AFTER LITERALLY CENTURIES OF TIRELESS RESEARCH AND ENDLESS DEBATE, WE REMAIN UNABLE TO ACCURATELY DESCRIBE AND PREDICT THE PROPERTIES OF WATER.**
> Richard Saykally

For example, if you thought that water only comes in three different forms – liquid water, steam and ice – you'd be wrong. Some models suggest there are two different liquid phases (see page 24) and as many as twenty different phases of ice. There are quite a lot of things about water we don't really know, but let's start with the things we do.

TIMELINE

6TH CENTURY BC	1781	1884
Greek philosopher Thales of Miletus calls water the source of all life	Composition of water revealed by Henry Cavendish	First proposal of water 'clusters'

WHY WATER IS ESSENTIAL TO LIFE

Water is everywhere. As the American chemist and water expert Richard Saykally is fond of reminding people, it is the third most abundant molecule in the Universe. It covers almost three quarters of our own planet's surface and, if you ever hear astronomers banging on about the hunt for water on Mars (see page 124), it's because they're interested in finding life elsewhere in the cosmos and water is really, really important for life. Liquid water especially. That's because it has some unique chemical and physical properties that make it ideally suited to hosting life and the chemical reactions that drive it.

First, liquid water is a fantastic solvent – it dissolves almost everything and many of the things it dissolves need to be dissolved in order to undergo reactions. That's what allows the other chemicals in our cells to react to form a working metabolism. It can also move chemicals around in a cell or a body and remains liquid at an unusually wide range of temperatures compared to other chemicals. You might think it

Water's contribution to climate change

Very recently, physicists at the Russian Academy of Science in Nizhny Novgorod got close to solving one of the mysteries that has for a long time plagued scientists studying the chemistry of our atmosphere. Water seems to absorb much more radiation than theoretical models based on its structure predict it ought to. The difference between predicted and actual values could be explained by the presence of dimers – doublet water molecules – floating around in the atmosphere, but no one has been able to prove they actually exist. To find these elusive dimers, Mikhail Tretyakov and his team went so far as to invent a brand new type of spectrometer for their experiments. Their results provided an absorption 'fingerprint' for water that was more clearly than ever associated with the suspected dimers and could help us understand exactly how water contributes to the infrared absorption spectrum of our atmosphere.

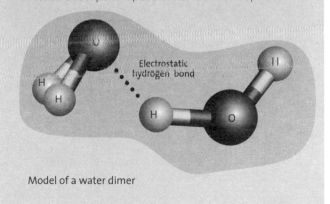

Model of a water dimer

1975
Boutron and Alben publish a model for ring-structured water molecules

2003
NASA spacecraft finds large amounts of water ice on Mars

2013
New evidence for water dimers in the Earth's atmosphere

obvious that water should freeze at 0 °C and boil at 100 °C, but you won't find much else that remains liquid across such gulfs. Ammonia, for instance, freezes at −78 °C and boils at −33 °C, and like ammonia, most other naturally occurring chemicals aren't even liquid at the sort of temperature at which life on Earth exists.

Life without water

We generally think of life as being dependent on water. But is that really true? Proteins, the molecules that form enzymes and structures like muscles in our bodies, were once thought to need water to keep their shape and carry out their many jobs. But in 2012, scientists at the University of Bristol, England, realized that myoglobin, the protein that binds oxygen in muscles, maintains its structure when it is deprived of water and, intriguingly, becomes extremely heat-resistant.

Water's other great asset is that it is denser when it is liquid than when it is solid – a result of the way that water molecules are packed together in ice and the reason that ice floats. Think what a mess the world would be in if icebergs sank.

WHAT ELSE WE KNOW ABOUT WATER

The water molecule is bent a bit like a boomerang and is very, very small, even compared to other common molecules like CO_2 and O_2, meaning you can really pack a lot of it into a small space. A one-litre bottle contains around 33 septillion water molecules – 33 followed by 24 zeroes. By some estimates, that's more than three times as many molecules as there are stars in the Universe. This close packing, plus the hydrogen bonds that attract the oxygen atoms of one molecule to the hydrogen atoms of others (see page 20), is what stops molecules from flying off and helps keep water as a liquid rather than a gas.

That's not to say that all the molecules in liquid water are stuck in the same place – far from it. Water is dynamic. The hydrogen bonds that hold water together break and reform trillions of times every second, so that there is hardly time for a cluster of molecules to form before it has already vanished. By contrast, the evaporation of a water molecule happens only very 'rarely', at just 100 million times a second from every square nanometre of water's surface.

<blockquote>
NOTHING IS EITHER GENERATED OR DESTROYED, SINCE A KIND OF PRIMARY ENTITY ALWAYS PERSISTS... THALES SAYS THE PERMANENT ENTITY IS WATER.

Aristotle, *Metaphysics*
</blockquote>

WHAT WE DON'T KNOW ABOUT WATER

We do know a lot about water, but there's also plenty we don't know. That rare evaporation event, for example, which requires the breaking of hydrogen bonds to free a molecule of water from the surface, is not very well understood. The fact that it doesn't happen very often doesn't help. And despite an array of cutting-edge techniques being used to probe the structure of water, those 'clusters' that appear to flicker in and out of existence aren't very well understood either. Even the idea of water clusters is questionable. If they exist so fleetingly, how can they make up something we can call a structure?

Hundreds of different models have been proposed to try to explain the structure of water, but none of them capture its behaviour in all its different forms and under a wide variety of different conditions. Research groups around the world, including Richard Saykally's at the Lawrence Berkeley National Laboratory in California, have been hard at work for decades trying to solve this remarkably complex problem. Saykally's group is using some of the most powerful and sophisticated spectroscopic techniques available, and resorting to quantum mechanical models, to explain the properties of this tiny molecule upon which all life relies.

The condensed idea
There's a lot going on below the surface

30 Origin of life

The origins of life on Earth have preoccupied scientists and thinkers from Charles Darwin to modern-day chemists. Everyone wants to know how life began, but the truth is it's a question that is difficult to answer definitively. There is, however, a point to all this pondering – to find the minimum criteria required to create artificial life in the lab.

Four billion years ago some chemicals got together and formed a prototype cell. Where this happened is a matter of debate – it may have occurred near the bottom of the ocean, in a warm, volcanic pool, foam-flecked mudflats or – if you believe in the theory of 'panspermia' – on another planet entirely. Location is everything, but for the time being it remains speculative.

Today, everything living emerges from other living things – animals give birth, plants make seeds, bacteria replicate and yeast bud. But the very first forms of life must have emerged from non-living stuff, as a result of ordinary chemicals bashing together and combining in the right way. The first cell would have been simple compared to a modern human or even bacterial cells. It was probably just a bag of chemicals that together comprised a very basic metabolism. Some sort of self-replicating molecule must also have been present so that information could be passed on to other, future cells. This might have made up a simple genetic code, but it would have been nothing as complicated as DNA (see page 140).

TIMELINE

1871	1924	1953
Darwin imagines life beginning in a 'warm little pond'	Oparin's *The Origin of Life* introduces the primordial soup theory	Publication of Stanley Miller's origins of life experiments

We can only guess at the molecules and conditions that started life on Earth; this guessing game is one that many chemists are keen to play. For not only does understanding first life teach us about our own origins, it inspires chemists who are trying to create new forms of life in the lab.

MILLER'S SOUP

You might well have heard of Stanley Miller and his famous experiments on the origins of life in the 1950s. Or at least, if you haven't heard of him, you'll probably have heard of his soup. Miller was the American chemist who many associate with the idea that life began in a primordial soup. Actually, his inspiration came from the lesser-known Aleksandr Oparin's 1924 book, *The Origin of Life*. His 'soup' was a concoction of methane, ammonia, hydrogen and water that he brewed in a flask in his lab at the University of Chicago. It was meant to represent the oxygen-less atmosphere of early Earth. To jolt the chemicals in the flask into action, he used an electrical spark to provide energy – simulating lightning in the early atmosphere.

> **IN THIS APPARATUS AN ATTEMPT WAS MADE TO DUPLICATE A PRIMITIVE ATMOSPHERE OF THE EARTH ...**
>
> Stanley Miller, writing in the journal *Science*, 1953

Miller's soup set-up yielded some of the first evidence that inorganic chemicals, with a little gentle persuasion, could come together to form organic molecules. For when Miller and his supervisor Harold Urey analysed the components of the soup a few days later, they found that amino acids – the building blocks of proteins – were present.

The primordial soup theory, however, is a little behind the times these days. While Miller's experiments are rightly regarded classic among fans and followers of chemistry, some doubt that he got his mix of ingredients correct, while others wonder whether lightning could really have provided the necessary constant source of energy to drive life forward from organic chemicals to cells. Unsurprisingly, there are a number of new theories

1986
RNA world hypothesis proposes that self-replicating RNA kickstarted evolution

2000
Discovery of Lost City hydrothermal vents

2011
Cambridge, England, team makes self-replicating RNA that can copy over 90 letters (bases) of code

The replication problem

At some point during evolution, cells must have adopted DNA as their information carrier, but before that, they could have used something simpler. RNA, a sort of single-stranded version of DNA, is such a molecule; but without the specialized copying machinery of modern cells, it would have had to reproduce itself. To do this, it must effectively have acted like an enzyme that could catalyse its own replication. This is all very well, of course, as long as you can find an RNA molecule that can replicate itself. But what if you can't? Doesn't that screw up your theory? Well, it does a bit. And that's been the problem with the theory for a long time – scientists have trawled through trillions of RNA molecules with different sequences looking for that special sequence that would code for self-replication but, as yet, they haven't found one that can do a decent job. Most 'self-replicators' can only copy portions of their own code, besides which copying accuracy is often pretty poor. The search continues ...

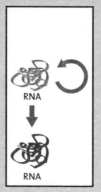

Modern world		RNA world
DNA	Information storage	
RNA	Information storage/ transmition	RNA
Protein	Function	RNA

knocking about concerning the exact whereabouts of these chemical beginnings.

THE LOST CITY

One contemporary theory suggests that life began in the deep ocean in a place called 'Lost City'. Sounds enticing, doesn't it? Lost City, in the Atlantic Ocean, was discovered in 2000, by a team of scientists led by Donna Blackman from the Scripps Institution of Oceanography in California. They were aboard the research vessel *Atlantis*, exploring an underwater mountain using a remote camera system, when they came across a field of hydrothermal vents – 30-metre tall chimneys spouting warm, alkaline water into the cool, dark ocean.

Though these vent systems exist elsewhere in the ocean and others had been discovered decades before, some chemists think that the Lost City vents provide the perfect conditions for the spawning of life on Earth. Here, hydrogen in the vent water and carbon dioxide in the seawater can meet and react, potentially forming organic chemicals. Not only that, the vent water – heated from below, by hot rocks under the ocean floor – provides a constant source of energy.

The other compelling aspect of the Lost City theory is that the acidity difference between the vent water and seawater mirrors the difference in acidity across the membrane of a cell. Could this be mere coincidence? It's not easy to test this theory in the deep ocean, but small-scale Lost-City-type reactors have been built in the lab.

BACK IN THE LAB

Not all chemists, however, study life's origins out of pure curiosity. Some are interested in figuring out the basic components that comprise life with a view to creating artificial life in the lab. We're not talking about creating artificial cows or cloning babies – this is more about using simple materials that can be used to make cell membranes. In real cells, such membranes are made of fatty molecules. The trick is to introduce some form of self-replicating system that allows these minimalist 'cells' to reproduce. There are some scientists who claim self-replicating protocells (see Protocells, above) are very close to being realized.

Protocells

In November 2013, Nobel Prize-winning biologist Jack Szostak and his team made a minimal cell or 'protocell' enclosed within a fatty envelope. Although it was simpler than even the simplest bacterium alive today, it did contain RNA capable of (roughly) copying itself. This copying was catalysed by magnesium ions. The chemical citrate also had to be added to stop the magnesium ions destroying the outer envelope. It may only be a matter of time before scientists make protocells that are entirely self-reproducing.

The question is: what are these protocells good for? Well, imagine if you were to design a self-replicating system that would just keep making more of itself as long as it got fed. What would you think about building into that system? The sensible answers, of course, are medicines and fuels. But why stop there? You might suggest anything of which a never-ending supply would be desirable – beer or strawberry shoelaces, for example. Scientists are already thinking outside the box – one suggestion is live, self-renewing paints.

The condensed idea
The stuff of life sprang from non-living matter

31 Astrochemistry

While the emptiness of space might suggest there's not much going on out there, it turns out there's more than enough to occupy chemists interested in the origin of life, not to mention the possibility of life elsewhere. So besides looking for the obvious – water on Mars, for example – what are they all up to?

Earth's atmosphere is rich with chemistry. It's chock full of molecules that are constantly crashing together and reacting. At sea level, every cubic centimetre contains around 10^{19} or 10,000,000,000,000,000,000 molecules. The vacuum of space is very different, by contrast. Each cubic centimetre of interstellar medium contains on average one single particle. Just one. This is equivalent to a bee buzzing around a city the size of Moscow.

Even if you only consider the scarcity of molecules, it seems pretty unlikely that two would ever meet and react. But there's also an energy problem to contend with. Earth's atmosphere is, on the whole, fairly warm, even if it might not feel like it on a crisp, winter morning in London or New York. In parts of the interstellar medium, however, the temperature can get down below a decidedly parky −260 °C. Things tend to move around pretty sluggishly in those kinds of conditions, which means any molecules that do meet may only be gently brushing past each other, and lack the necessary energy to react with each other. Given this particular set of unlikely circumstances, it's surprising any chemistry ever gets done at all. It rather begs the question why chemists are interested in what goes on in space.

TIMELINE

13.8 BILLION YEARS AGO	400,000 YEARS AFTER THE BIG BANG	1937
The Big Bang	The first molecules were formed – chemistry begins!	First interstellar molecules identified

HOTSPOTS

Despite the apparent paucity of actual chemistry, there are plenty of chemists interested in studying whatever is up there in space and for a few good reasons, too. The chemistry of space can tell us about how the Universe started, where the chemical elements for life came from and whether life might exist anywhere but our own planet. But before we can even consider the more complex chemistry of biological reactions, we need to think more about the conditions in space, what molecules are present and how they set the scene for basic reactions to occur.

Just looking at the average conditions in space doesn't tell us much about what it's like in any particular spot. In places, it might be sparse and cold, but space is so massive that conditions can vary wildly. The interstellar medium, which fills the space between stars, is not just a uniform sea of gas particles. There are cold, dense molecular clouds containing hydrogen, but there are also superhot spots around stellar explosions.

> **WE HAVE ABOLISHED SPACE HERE ON THE LITTLE EARTH; WE CAN NEVER ABOLISH THE SPACE THAT YAWNS BETWEEN THE STARS.**
> Arthur C Clarke,
> *in Profiles of Our Future*

Most (99 percent) of the interstellar medium is made up of gases – hydrogen, makes up more than two thirds by mass, and helium accounts for most of the rest of it. The amounts of carbon, nitrogen, oxygen and other elements are tiny in comparison. The other 1 percent is a component that may sound curious to those who have read Philip Pullman's *His Dark Materials* trilogy: dust. This dust does not resemble the dust you might wipe from your window sill, or even – for the benefit of Pullman fans – fictional, conscious particles.

DUST

Interstellar dust is made up of small grains that contain substances such as silicates, metals and graphite. What's important about these dust particles is that they give lone molecules floating through the vast emptiness of space

1987	2009	2013
Detection of acetone in interstellar medium	Total molecules detected in interstellar medium rises above 150	Titanium dioxide identified in space

Life on Mars

Our nearest neighbour in the Solar System, Mars, has always attracted attention from scientists looking for life elsewhere in the Universe. The presence of water, which astrobiologists consider essential to life, was at first taken as a sign that life could actually exist there. Since then, it has become apparent that water on Mars is largely trapped as ice below the surface or clinging to soil particles. Theoretically, a thirsty astronaut could heat up a few handfuls of Martian soil to get a gulp of water. In 2014, images were published in the Solar System science journal *Icarus* that showed what looked suspiciously like gullies on the surface, leading some to suggest that water once flowed over the Red Planet. But there's no evidence that water on Mars – in whatever form – once sustained life, or that it does today.

somewhere to stick around. And if they stick around for long enough they might eventually meet another molecule they can react with. Some grains are encased in ice (water ice), so ice chemistry is key to understanding what might happen on these grains. Other elements in dust particles may offer catalytic services, helping the rare reactions chug along. Where energy levels are low, reactions may also be helped along by UV radiation in starlight, cosmic rays and X-rays, while there are some reactions that don't require energy at all.

In 2013, astronomers making radio observations of the distant sky with the Submillimeter Array telescope in Hawaii discovered signs of titanium dioxide in dust particles around the very bright supergiant star VY Canis Majoris. Titanium dioxide is the same chemical that is used in sunscreen and to make the pigment in white paint. They suggested that, in space dust, the chemical might be important for catalysing reactions that form larger, more complex molecules.

SEEDING LIFE

Larger molecules are, however, something of a rarity in space, as far as we know. It is less than 80 years since the first interstellar molecules – the radicals CH˙, CN˙ and CH⁺ – were identified. Since then, about another 180 have been confirmed, with most having six atoms or less. Acetone – $(CH_3)_2CO$ – having ten atoms, is one of the larger molecules and was first detected in 1987. Large carbon-containing molecules, like the polycyclic aromatic hydrocarbons (PAHs), are the ones that astrochemists are really interested in, because they

PAHs

Polycyclic aromatic hydrocarbons (PAHs) are a diverse group of molecules, all containing benzene ring structures. On Earth, they are products of incomplete burning and pop up in burnt toast and barbecued meat, as well as in fumes from cars. They have been detected all over the Universe since the mid 1990s, including in early, star-forming regions, although their presence hasn't been directly confirmed.

Anthanthrene
$C_{22}H_{12}$

Napthalene
$C_{10}H_8$

Pyrene
$C_{16}H_{10}$

Chrysene
$C_{18}H_{12}$

might tell them something about how organic molecules first formed. PAHs and other organic molecules are often linked to theories of the origins of life in which they are imagined to have seeded life on Earth. Amino acids have also been detected, but not confirmed.

Astrochemists don't just look for the signatures of interesting molecules. They have other tools in their toolbox. They can simulate what might be happening in space in their own labs. Using vacuum chambers, for example, it's possible to recreate small pockets of the vast interstellar 'emptiness', which we know is not entirely empty, and try to work out how reactions could take place there. Together with modelling, this approach predicts molecules and reactions that might later be confirmed as technology progresses. New, powerful telescopes like the Atacama Large Millimeter Array, in Chile's Atacama Desert, should help chemists to prove, or disprove, some of their furthest flung theories.

The condensed idea
Chemistry with telescopes

32 Proteins

Protein is supposed to make up a key part of our diet, but do we really know why? What is protein actually doing in our bodies? A lot more than we might give it credit for, as it happens. Protein is the multipurpose molecule for hire – it comes in an unimaginable number of different forms and each one is uniquely suited to its task.

From the strength and elasticity of spider silk to the ability of antibodies to defend us against disease, the extraordinary diversity within the structures of proteins translates into a wealth of different functions. While we all know that our muscles are built from proteins, we sometimes forget that this family of molecules is responsible for much of the hard graft that goes on inside living things. They are often called the 'workhorses' of the cell. But what are proteins?

BEADS ON A STRING

Proteins are chains of amino acids joined together by peptide bonds. Imagine a string of coloured beads where each different colour represents a different amino acid. There are around 20 different colours or amino acids that are found in nature. Those that your body makes are called non-essential amino acids, while those that you need to get from your food are called essential amino acids (see Essential and non-essential amino acids, page 131).

Not all amino acids are made by living organisms. A meteorite that crashed near Murchison, Australia, in 1969, was carrying at least 75 different amino

TIMELINE

1850	1955	1958
First synthesis of an amino acid (alanine) by Adolf Strecker	Amino acid sequence of insulin by Frederick Sanger	Kendrew and Perutz produce first high-resolution protein structure (myoglobin) by X-ray crystallography

acids. Only the decade before, Stanley Miller's experiments on the origin of life (see page 121) had proved that amino acids could be made from simple, inorganic molecules under conditions like those on a four-billion-year-old Earth.

Each amino acid is based on a universal structure – the most general form of which is $RCH(NH_2)COOH$. This includes a central carbon atom joining together a NH_2 (amine), a COOH (carboxylic acid) and a hydrogen atom. The 'R' group joined to the central carbon is the part that gives the amino acid its unique properties. Spider silk, for example, contains lots of glycine, the smallest and simplest amino acid, which has only an extra hydrogen as its R group. Glycine is thought to contribute to the elasticity of the fibres.

The order in which the beads are arranged on the protein string is called the protein's primary structure – its amino acid sequence. So, like DNA, protein can be 'sequenced'. Depending on the type of silk and how it is used, spider silk proteins have slightly different amino acid sequences. However, it's thought that about 90 percent of each sequence is made up of repeated blocks of between ten and 50 amino acids.

> **WHEN I SAW THE ALPHA-HELIX AND SAW WHAT A BEAUTIFUL, ELEGANT STRUCTURE IT WAS, I WAS THUNDERSTRUCK.**
>
> Max Perutz, upon discovering alpha-helix structure of haemoglobin

SUPER STRUCTURES

The higher levels of structure in proteins are formed from the folding and coiling (secondary structure) of the amino acid chains into their overall three-dimensional shape (tertiary structure). Some secondary 'motifs' crop up again and again: to return once more to the example of spider silk, the strong silk that common orb-weaving spiders use to build the frames of their webs is made up of chains that are held together in sheets by extensive hydrogen bonding (see page 20). This so-called ß-sheet motif is also found in keratin, another structural protein that forms part of your skin, hair and nails.

1988	2009
Chymosin protein made by GM yeast approved for use in food	Nobel Prize for Chemistry awarded for work on protein assembly reactions

Linking together amino acids

The cell machinery responsible for threading amino acid beads onto a protein string is the ribosome. Its job is to form the peptide bonds linking each bead – a link is formed as the carboxyl group of one amino acid reacts with the amino group of the next, releasing a molecule of water. The ribosome is capable of linking around 20 new amino acids together every second, using the instructions provided by the DNA code. This rapid work rate means that the chemical reaction that forms the bonds has been difficult to study. But having already used X-ray crystallography (see page 88) to see the structure of the ribosome, the American chemist Thomas Steitz managed it. He went on to crystallize the ribosome at various moments of the linking reaction to produce three-dimensional structures that revealed the steps in detail and enabled the important atoms to be pinpointed. In 2009, Steitz was awarded the Nobel Prize in Chemistry for his work.

Glycine linked to alanine to form dipeptide glycylalanine

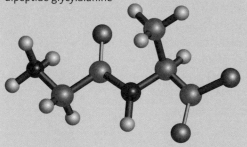

An even more common motif is the springlike alpha-helix structure found in haemoglobin – the oxygen-carrying component of blood – and the muscle protein myoglobin.

In spider silk, it is the beta sheets that are thought to be responsible for the protein fibres' strength, which is on a par with steel. (It's worth noting that this incredible strength is combined with elasticity greater than nylon and toughness greater than man-made Kevlar, used in bullet-proof vests.) The fibres have given inspiration to several firms now attempting to produce artificial spider silk. One, made by Kraig Biocraft Laboratories, is a spider-silk-like fibre called Monster Silk spun by genetically modified silk worms. The company doesn't just intend to copy natural silk; it intends to improve on it, for instance, by incorporating antibacterial functions.

MULTIPLE ROLES

Proteins don't just build structures, they control and enable much of what is happening inside the cell. By some estimates, a typical animal cell is about 20 percent protein and contains thousands of different types of proteins. This diversity of forms is not so hard to imagine once you realize that there are more than three million possible combinations of beads on a protein string, just five amino acids long, and most are much, much longer. But even when proteins aren't building structures, their shape remains crucial.

One of the most important roles that proteins perform in the cell is as biological catalysts – enzymes (see page 132) – that control chemical reaction rates. Here, protein structure and three-dimensional shape are key, because they determine how the enzyme interacts with the molecules involved in the reaction. Biological catalysts are often highly specific to the reactions they shepherd, and more so than the chemical catalysts used to speed up reactions in industry.

Essential and non-essential amino acids

In adult humans, the essential amino acids are phenylalanine, valine, threonine, tryptophan, isoleucine, methionine, leucine, lysine and histidine. Non-essential amino acids, which have to be absorbed from food, are generally alanine, arginine, aspartic acid, cysteine, glutamic acid, glutamine, glycine, proline, serine, tyrosine, asparagine and selenocysteine. Some people's bodies, however, can't make all of these non-essential amino acids and so they need to take food supplements to obtain them.

Protein structure is also vital to the immunoglobulin molecules – antibodies – that our immune systems deploy to fight disease. When you get a particular strain of flu, your body produces antibodies against it that stop you succumbing to that particular strain in the future. The antibodies are protein-based immunoglobulin molecules that recognize and bind specifically to a portion of the flu virus, and their recognition is based on their structure. Through rearrangements in the genes of antibody-producing cells, our bodies are able to create protein structures to deal with millions of different invaders.

Unfortunately, the importance of protein structure is never more apparent than when something goes awry. Parkinson's disease is the result of misfolded proteins in nerve cells. Scientists are still trying to understand whether malformed proteins are also at the root of other devastating diseases, such as Alzheimer's.

The condensed idea
Function follows form

33 Enzyme action

As biological catalysts, enzymes drive reactions from the metabolic processes of our own bodies to the reactions that allow viruses to multiply inside our cells. Two models of enzyme action have dominated our thinking on how enzymes work during the last century. Both models try to explain how each enzyme is specific to the reaction that it catalyses.

The German biochemist Hermann Emil Fischer seems to have had a curious obsession with hot drinks, focusing his interest on the purine chemicals in tea, coffee and cocoa. At some point, he added sugars to the mix – and milk, in the form of lactose. In a roundabout way, this led him to the study of enzymes. In 1894, he proved that the hydrolysis reaction that splits lactose into its two component sugars can be catalysed by an enzyme and, in that very same year, published a paper that outlined a theory for how enzymes work.

LOCK AND KEY

Enzymes are the biological catalysts (see page 48) that drive the reactions in all living things. Fischer's 'lock and key' theory of enzyme action was based on the observation that one of his precious sugars came in two slightly different structural forms (isomers) whose hydrolysis reactions were catalysed by two different enzymes from natural sources. The reaction of the 'alpha' version only worked with an enzyme from yeast, while the reaction of the 'beta' version only worked with an enzyme from almonds. Even though the two sugars contained all the same atoms, joined up for the most part in the same ways, they didn't both fit the same enzymes. Fischer regarded the two forms

TIMELINE

1894	1926
Hermann Emil Fischer proposes 'lock and key' model of enzyme action	First crystallization of enzyme (urease) by James Sumner

of the sugar as keys that only fit the correct locks.

Expanding this theory to enzymes and their substrates (the 'keys') more generally, Fischer developed the first model of enzyme action that could explain a crucial characteristic of enzymes: their specificity. It wasn't until decades after Fischer's death that his model was overturned – but in the meantime, there was other work to be done on enzymes.

PROVE THEM WRONG

One fact that did not become apparent to Fischer was that all enzymes share the same molecular descent – they are proteins, made from amino acids (see page 128). This was clear to James Sumner, another charismatic chemist, but he had a hard time proving it. Sumner was a stubborn character – despite having his left arm amputated above the elbow after a childhood hunting accident, he resolved to excel in sports and eventually won the Cornell Faculty Tennis Club prize. His stubbornness obviously extended to his research because after several people had advised him that it would be foolish

The active site

The active site of an enzyme is the portion that holds the substrate and where the reaction between enzyme and substrate occurs. It may be formed from just a few amino acids. Anything that changes the structure of the active site alters the fit and makes it less likely that the reaction will occur. For example, an increase or decrease in pH affects the number of hydrogen ions floating around (see page 44). These hydrogen ions interact with groups on the amino acids in the active site and alter the structure. Any molecule that binds to an enzyme in such a way that it directly blocks the active site is called a competitive inhibitor, since it 'competes' with the substrate. Molecules that bind elsewhere but still change the structure enough to make the enzyme useless are called non-competitive inhibitors. Genetic changes can also affect enzyme action, especially if they translate into changes to the amino acids in the active site. For example, in Gaucher disease, mutations that affect the active site of an enzyme called glucocerebrosidase mean that its substrate accumulates in the organs. However, it's possible to replace the defective enzyme – globally around 10,000 people with Gaucher disease are receiving enzyme replacement therapy.

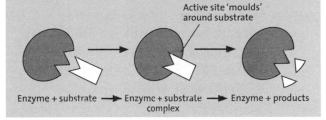

Active site 'moulds' around substrate

Enzyme + substrate → Enzyme + substrate complex → Enzyme + products

to try to isolate an enzyme, he went ahead and tried to do it anyway – and it took nine years.

In 1926, Sumner became the first person to succeed at crystallizing an enzyme, isolating urease from jack beans. (Urease is also the enzyme that allows *Helicobacter pylori* to thrive in the human stomach, where it causes stomach ulcers. The enzyme breaks down urea to raise the pH and make the surroundings more comfortable.) When no one believed Sumner's claim that urease was a protein, he went on a mission to prove them wrong, publishing ten papers on the subject – just to make sure the fact was indisputable. It did also help Sumner's cause that he was awarded the Nobel Prize in Chemistry.

A BETTER FIT

At that time, the 'lock and key' model was still the preferred way of thinking about enzyme action. If urease was the lock, then urea was the key. Then, in the 1950s, the American biochemist Daniel Koshland revised Fischer's ageing model. His 'induced fit' model is the one that survives today. Koshland adjusted the rather rigid lock of Fischer's theory to accommodate the fact that enzymes are made up of protein chains, which have a more flexible structure.

> **A NUMBER OF PERSONS ADVISED ME THAT MY ATTEMPT TO ISOLATE AN ENZYME WAS FOOLISH, BUT THIS ADVICE MADE ME FEEL ALL THE MORE CERTAIN THAT IF SUCCESSFUL THE QUEST WOULD BE WORTHWHILE.**
> *James Sumner*

Proteins and enzymes can be affected by conditions such as temperature – above body temperature the activity of human enzymes quickly falls off a cliff – and the presence of other molecules. Koshland realized that when a substrate molecule meets its very specific enzyme, it causes a change in the shape of the enzyme that results in a tighter fit. Hence 'induced fit'. This occurs in the area of the active site, the small portion of the enzyme that forms Fischer's lock. So urea doesn't slip seamlessly into urease. It's more as if it were shuffling about on a beanbag to get comfortable.

The induced fit model has also come to have wider relevance in understanding binding and recognition processes in biology. It is important, for example, in understanding how hormones bind to their receptors and how certain drugs work. HIV drugs like nevirapine and efavirenz work by binding to the enzyme called reverse transcriptase, which the virus uses to make DNA inside a human cell so that it can replicate. The drugs bind to a site next to the active site of the enzyme, causing a change in its structure and stopping the enzyme from doing its job. So the virus can't make new DNA and can't replicate.

Enzymes in industry

Enzymes are used in a wide range of different industries to facilitate reactions. Biological washing powders contain enzymes that break down the substances in stains, saving on the energy needed to get clothes clean. The food and drinks industries use enzymes to convert one type of sugar into another. The trouble is that because enzymes are proteins, they only work under a narrow range of conditions, so temperature, pressure and pH, for example, must be strictly controlled.

Both models of enzyme action are taught in schools and are a great example of how scientific thinking evolves as new evidence comes to light. Daniel Koshland's revision was based partly on evidence relating the flexibility of protein structure and various anomalies in patterns, which led him to believe something wasn't quite right with the prevailing theory. But having the utmost respect for Fischer, who has become known as the Father of Biochemistry, Koshland always maintained that he only built on the great man's work. Touchingly, he wrote, 'It is said that each scientist stands on the shoulders of the giants who have gone before him. There can be no more honoured place than to stand on the shoulders of Emil Fischer.'

The condensed idea
Natural catalysts

34 Sugars

Sugars are nature's fuels, and along with proteins and enzymes, some of the most important biomolecules. They give your muscles power to run and your brain the energt to think. They even sew together your DNA. But they can also make you fat and allow invading viruses to enter your cells.

If you order a takeaway pizza on a Friday night, you might decide to go for a run on Saturday morning to burn it off. When we say that we 'burn off' food, we're usually referring to the reaction that our body uses to break down sugar to provide us with energy. Like coal, sugar is a fuel and it needs oxygen to burn efficiently to produce energy, carbon dioxide and water. While we have to eat to get our sugars, plants make their own through the photosynthesis reaction (see page 148), which is why most of the sugar in our food comes from plants.

But sugar is not just nature's fuel. In the knowledge that coal, oil and gas are running out, humans are becoming increasingly caught up in schemes to extract the energy from plants on a massive scale. The biofuels industry promises to deliver renewable energy from sugars, and complex sugars like starch and cellulose, stored in crops and plant waste – though it must then compete with food producers for land.

Sugars have other uses besides energy. In ribose, they form an integral part of the DNA and RNA molecules that carry the genetic code. They combine with proteins to form receptors on cells – for example, allowing viruses to enter – and can relay messages between distant cells, acting like hormones. And

TIMELINE

1747	1802	1888
German chemist Andreas Marggraf extracts crystals from sugar-beet juice and compares to crystals in sugar cane	First sugar-beet refinery begins operations	Emil Fischer discovers the link between glucose, fructose and mannose

what's more, rather surprisingly, plants use sugars to tell the time.

RIGHT ON THE '-OSE'

The sugar you spoon into your tea or coffee is sucrose, the same form that plants store and we extract from sugar cane or sugar beet. But there are many different chemical forms of sugar. You can often pick out sugars in a list of ingredients by their giveaway 'ose' suffix: glucose, fructose, sucrose, lactose. Chemically, they're all carbohydrates – hydrated carbons. Some are short chains, others are ring-shaped, but very basically, they all contain carbon atoms with a double-bonded oxygen atom (see Sugars and stereoisomers, right). Emil Fischer, the Nobel Prize-winning chemist who did pioneering work on sugars, was the first to understand the link between glucose, fructose and mannose in 1888.

The less recognizable forms of sugar are those that are long chains of sugars strung together to make polymers or polysaccharides. One example is maltodextrin, a glucose polymer that comes from maize or

Sugars and stereoisomers

This image below shows two versions of glyceraldehyde – a simple (monosaccharide) sugar. Like glucose, it contains one aldehyde (–CHO) group. All sugars contain ketone or aldehyde groups. In a ketone group, the oxygen is attached to a carbon that is bonded to two other carbon-containing groups, whereas, in an aldehyde group, the carbon bearing the double-bonded oxygen uses up one of its two other bonds on a hydrogen atom. You can see that the two structures look very similar, except 'L' glyceraldehyde appears to have its OH and H groups attached the opposite way around to 'D' glyceraldehyde. There is no way of rotating L to make it the same as D. This is because the two molecules are stereoisomers – although their atoms and bonds are identical, the overall 3-D arrangement is different. One special type of stereoisomer is an enantiomer, where two stereoisomers are mirror images (see page 72). The convention for drawing stereoisomers in 2-D was developed by Emil Fischer in 1891, while he was working on sugars.

Fischer projection formula

D-Glyceraldehyde

CHO
H —— OH
CH₂OH

CHO
H ····· OH
CH₂OH

l-Glyceraldehyde

CHO
HO —— H
CH₂OH

CHO
HO ····· H
CH₂OH

1892	1902	2014
Fischer establishes 3-D arrangements of 16 hexose sugars	Nobel Prize for Chemistry awarded to Fischer for work on sugar and DNA bases	Chemists announce a wearable blood sugar-sensing device

Sugar sensing

Being able to sense sugar levels in our blood is important, medically, for those with diabetes or trying to lose weight. In 2014, chemists and technologists at a new company Glucovation announced they had combined their expertise to develop the first wearable sensor for blood sugar that could track glucose levels all day. Instead of sticking in a new needle each time, diabetes (and health fanatics) would be able to insert one every week and monitor glucose levels on their smartphones.

wheat and is added to the energy powders and gels used by athletes. Scientists are also developing biodegradable batteries that use maltodextrin as an energy source. As in nature, the batteries use enzymes – as opposed to expensive, catalytic metals used in traditional batteries – to drive the reactions that produce the energy.

ONE WAY OR ANOTHER

As far as humans are concerned, perhaps the most important form of sugar is glucose – a simple monosaccharide, which consists of just one type of sugar. Sucrose, by contrast, is a disaccharide, because it's formed from glucose and fructose joined together by a glycosidic bond. The enzyme-driven process that we use to extract energy from the sugar in our food is a complex, multistep reaction that supplies living cells with energy.

Here's the reaction:

$$C_6H_{12}O_6 + 6O_2 \rightarrow 6CO_2 + 6H_2O$$

glucose + oxygen → carbon dioxide + water (+ energy)

It's actually a bit more complicated than that, but this summary reaction at least tells us which are the initial reactants and end products. The oxygen part is important because, without it, glucose doesn't burn so efficiently and is converted to lactic acid, the chemical produced by fermenting yeast and is also associated with fatigue during exercise. Although the body can get energy by making lactic acid, the return is much lower.

There is a lot of interest in sports science in understanding how these two systems – aerobic and anaerobic – overlap during, for example, a track race.

For instance, runners in both the 400-m and 800-m events use aerobic energy produced through the normal route, but because the muscles can't get enough oxygen to produce the power that is required, they must also produce energy anaerobically. The contribution of the aerobic route only starts to overtake that of the anaerobic route after 30 seconds or more of running, so an elite 400-m runner finishing in 45 seconds must use mostly lactic acid, while an 800-m runner's energy comes mostly from the 'normal' glucose-processing system.

SUGAR O'CLOCK

Though sugar is an important source of energy, we're all well aware that our sugar levels need to be finely balanced. Excess glucose is stored in the liver and muscles as the polysaccharide glycogen, which is fine if you're the aforementioned elite 400-m runner, who's going to burn it all off. However, if too much sugar is hanging around, the body will transform it into fat and pack it into fat cells as an energy-rich back-up fuel, should you suddenly decide to start marathon training. Meanwhile, the brain only performs well on glucose, which could be seen as a good excuse for ploughing through the odd cake during a difficult afternoon at work.

Still wondering how plants use sugar to tell the time? Well, in 2013, researchers at the Universities of York and Cambridge, England, discovered that plants use the build-up of sugar during the day to set their circadian clocks. When the Sun comes up in the morning, they start photosynthesizing. The sugar accumulates and eventually reaches a certain threshold level that indicates to the plant that dawn has arrived. The researchers showed that stopping plants from photosynthesizing also screwed up their circadian rhythms, but giving them sucrose helped them reset their clocks.

... SUGAR, NATURE'S FIRST ORGANO-CHEMICAL PRODUCT, FROM WHICH ALL OTHER CONSTITUENTS OF THE PLANT AND ANIMAL BODY ARE REFORMED.
Emil Fischer

The condensed idea
Fuel and foe

35 DNA

James Watson and Francis Crick are often painted as the main protagonists in the story of DNA. But we shouldn't forget that some of the early research on the chemical contents of cells was vital to our discovery of the genetic material – and arguably more interesting.

Any ordinary person's stomach would turn at the idea of sorting through other people's pus-soaked bandages. But Friedrich Miescher was no ordinary person. He was the kind of person who was interested enough in the contents of pus to devote a large portion of his working life to studying them. He was also the kind of person who was prepared to rinse out pigs' stomachs, and embark on late-night fishing trips in order to get his hands on cold salmon sperm.

Miescher's aim was to obtain the purest possible samples of a substance he called nuclein. Despite training as a medical doctor, the Swiss scientist had joined the biochemistry lab of Felix Hoppe-Seyler at Tübingen University, Germany, in 1868, and become fascinated with the chemical components of cells. That fascination never went away, and while Miescher might not be the most familiar of the scientists associated with studying DNA – James Watson and Francis Crick, who proposed its structure, are much better known – his discoveries were surely some of the most important.

PUS AND PIGS' STOMACHS

Miescher's supervisor, Hoppe-Seyler, was interested in blood, and so it was that Miescher's early investigations focused on white blood cells, which

TIMELINE

1869	1952	1953
Friedrich Miescher extracts 'nuclein' (DNA) from white blood cells	DNA confirmed as genetic material	Double helix structure of DNA published

he found he could collect in large quantities from the pus that soaked into wound dressings. He got these fresh from a nearby surgery. By chance, cotton wool had not long been invented and proved an excellent material for absorbing the pus. At this point, Miescher had no grand ideas about identifying the material responsible for inheritance – he was just hoping to learn more about the chemicals present within cells.

At some point in his studies, Miescher came across a precipitate, which, although it behaved a bit like protein, he could not identify as any protein already known. It seemed to come from the nucleus, the mass at the centre of the cell. As his interest in the material inside the nuclei grew, he tried various strategies to isolate it. This was where the pigs' stomachs came in. Pigs' stomachs are a good source of pepsin, a protein-digesting enzyme that Miescher used to break down most of the other contents of the cells. To get the pepsin, he swilled hydrochloric acid around in the stomachs. Using the pepsin, he finally got a fairly pure sample of a grey substance that he called 'nuclein' – it contained what we now know as DNA.

> **DNA AND RNA HAVE BEEN AROUND FOR AT LEAST SEVERAL BILLION YEARS. ALL THAT TIME THE DOUBLE HELIX HAS BEEN THERE, AND ACTIVE, AND YET WE ARE THE FIRST CREATURES ON EARTH TO BECOME AWARE OF ITS EXISTENCE.**
> Francis Crick

Miescher was convinced enough that this nuclein was crucial to understanding the chemistry of life that he persevered with its elemental analysis, reacting it with different chemicals and weighing the products to try to work out what it consisted of. One element that seemed to be present in unusually high quantities was phosphorus and it was this that persuaded Miescher that he must have found an entirely new organic molecule. He even measured the quantities of nuclein present at different stages of a cell's life and discovered that levels peaked just before division. This should have been a massive clue as to its role in information transfer and Miescher did indeed consider that nuclein could be involved in inheritance. But he ultimately dismissed the idea because he couldn't believe that one chemical could contain all the information to encode so many diverse forms of life –

1972	**1985**	**2001**	**2010**
Paul Berg assembles DNA molecules, using genes from different organisms	Polymerase chain reaction (PCR), a method for making millions of copies of DNA	Human Genome Project completed	Craig Venter creates a synthetic genome and inserts it into a cell

Miescher went on to find the substance in the sperm of the salmon that he plucked from the Rhine and later in carp, frog and chicken semen.

PIECING THE PUZZLE TOGETHER

One of the problems with Miescher's work on nuclein was that it went against the assumption of many scientists that protein was the inherited material. At the beginning of the 20th century, attention again turned to protein. By this time, the components of nuclein, or DNA, had already been uncovered: phosphoric acid (forming the DNA 'backbone' accounting for Miescher's phosphorus), sugar and the five bases that we now know make up the genetic code. But protein theories just seemed more convincing. The 20 amino acids in proteins offered greater chemical diversity and could therefore account for the great diversity of life.

The genetic code

Deoxyribonucleic acid (DNA) is made up of two chains of nucleic acids twisted together like fibres in a rope. The nucleic acids chains are repeated units where each unit is formed from the combination of a base, a sugar and a phosphate group. Two chains are held together by hydrogen bonds (see page 20) between the bases, whose sequence forms the genetic code. However, the base adenine only usually bonds to thymine (A–T), while the base cytosine only usually bonds to guanine (C–G). The code is copied when, in cell division, the hydrogen bonds break and the two strands separate to form templates for the creation of new strands, made by enzymes within the cell. To make proteins, the machinery of the cell reads the sequence of bases, translating trios of bases (codons) into single amino acids that are added to growing protein strings (see page 128). There are several different three-base sequences that code for each amino acid. So, serine, for instance, might be added as the result of the translating machinery reading a TCT, TCC, TCA or TCG codon.

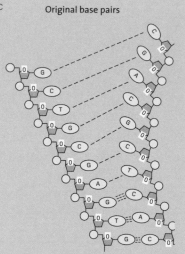

Original base pairs

The secrets of DNA began to unfold in the 1950s, when, all within a couple of years, studies confirmed that it was the genetic material transferred when a virus infected a bacterium, and the double helix structure was proposed by James Watson and Francis Crick. The contribution of a bright, young chemist and X-ray crystallographer (see page 88) named Rosalind Franklin to the structure published in *Nature* has too often been overlooked. It was Franklin, working at King's College, London, who took the photographs of DNA that inspired the structure. Her colleague Maurice Wilkins had shown the images to Watson without asking her. Franklin, meanwhile, was not even allowed to eat her lunch in the same room as the male scientists in her lab and had it not been for the support of her mother and aunt, her father would have refused to pay for her degree because he didn't believe women should have a university education.

Nucleotides

The combination of each DNA base with its sugar and a phosphate group is called a nucleotide. Technically, in DNA, nucleotides are deoxyribonucleotides because the sugar they contain is deoxyribose. In RNA – the single stranded version that cells use to translate the DNA code into proteins – the sugar is ribose, so the nucleotide is called a ribonucleotide. Oligonucleotides are short chains of nucleotides joined together.

THE DNA DICTIONARY

Finding the structure of DNA didn't solve the mystery entirely, however. More than half a century after Miescher died from tuberculosis, aged 51, it still wasn't clear how the diversity of life could emerge from nucleic acids. But after Watson, Crick and Wilkins received their Nobel Prize in 1962, another was awarded, in 1968, to Robert Holley, Har Gobind Khorana and Marshall Nirenberg for cracking the genetic code – they showed how the chemical structure of DNA translates into the chemical structure and complexity of proteins. Even now, despite the sequencing of the entire human genome, we are still trying to work out what much of it means.

The condensed idea
Chemical copies of life's code

36 Biosynthesis

So many of the chemicals that we use today, including life-saving antibiotics and dyes used to colour our clothes, are borrowed from other species. These chemicals can be extracted directly, but if the biosynthetic routes can be traced, they can also be made in the lab – through chemistry or with the help of surrogate organisms, such as yeast.

In January 2002, a team of South Korean scientists headed to the Yuseong forest in Daejeon, South Korea, to collect soil samples from the forest floor. Working among the pine trees, they took samples from the topsoil and from the loose earth around plant roots. They weren't interested in the soil itself, but in the millions of bugs living in it. They were looking for bacteria that produced interesting compounds that were new to science.

Back at the lab, they extracted the DNA from these microbes, along with other bugs from the Jindong Valley forest, and eventually inserted random pieces of their DNA into *Escherichia coli*. When they encouraged these bacterial clones to grow, they noticed something odd: some of them were purple. This wasn't what they'd been looking for. They'd been hoping they might find bugs that produced antimicrobial compounds – ones with potential as drugs – a bit like Alexander Fleming did when he discovered penicillin, the first antibiotic, in *Penicillium* mould.

After purifying the purple pigments and subjecting them to various spectral analyses – including mass spectrometry and NMR (see page 84) – the researchers realized they weren't even new pigments. Weirdly, they were

TIMELINE

1897	1909	1928
Ernest Duchesne discovers that *Penicillium* mould kills bacteria	Chemical analysis of Tyrian purple dye pigment	Penicillin discovered (or rediscovered) by Alexander Fleming

indigo blue and the red dye indirubin, two compounds usually made by plants, which were apparently being made here by bacteria.

NATURAL PRODUCTS

This is an interesting example of biosynthesis – the synthesis of natural products – because it shows how species from completely different branches of the evolutionary tree can end up making the exact same compounds. The Australian dog whelk and many other marine molluscs also make a compound related to indigo blue known as Tyrian purple, which like indigo has been used since ancient times to dye clothes.

Biosynthesis refers to any biochemical route – probably involving a number of different reactions and enzymes – that a living thing uses to make a chemical. Chemists, however, usually refer to biosynthetic routes that result in useful and or commercially exploitable natural products when they talk about biosynthesis. Such was the case for Fleming's penicillin, obviously, as well as indigo blue and Tyrian purple. Although there are now synthetic indigos and purples, Tyrian purple is still extracted from snails at great cost. It takes 10,000 *Purpura lapillus* snails to produce one gram of Tyrian purple, which in 2013 cost a staggering €2,440. There are plenty of other examples, too. Cheesemakers have been reliant on natural products from *Penicillium roqueforti* – a relative of the penicillin-producing mould – for centuries, in making blue cheeses such as Roquefort and Stilton.

> **NATURE BEING A SOPHISTICATED, VERSATILE AND ENERGETIC COMBINATORIAL CHEMIST ... YIELDS BY INFINITE NUMBER OF DIFFERENT AND UNPREDICTABLE WAYS, A SERIES OF EXOTIC AND EFFECTIVE STRUCTURES ...**
>
> János Bérdy, IVAX Drug Research Institute, Budapest, Hungary

Most natural products, from antibiotics to dyes, are chemicals called secondary metabolites. While primary metabolites are the sorts of chemicals that organisms need to sustain life – like proteins and nucleic acids – secondary metabolites are those that seem to be of no obvious use to the

1942	2005	2013
First patient treated with penicillin – Anne Miller treated for blood poisoning	Number of known natural products reaches approximately one million	Sanofi launches production of the antimalarial drug, artemisinin

How did we get from bread mould to penicillin?

The species of mould from which Alexander Fleming originally extracted penicillin was called *Penicillium notatum*. It's a type of mould that grows quite happily on bread in your kitchen. Fleming and his colleagues tried for years to get it to produce enough of the antibiotic to make it useful for treating patients. It was partly a purification problem, but they eventually realized that this particular species just didn't produce enough. They started looking for other similar strains that would give a better yield and, eventually, they found one growing on a cantaloupe

Structure of penicillin
(R is variable)

melon – *Penicillium chrysogenum*. After subjecting it to various mutation-inducing treatments, such as X-rays, they had on their hands a species that could produce 1,000 times the amount, and it is still used today.

organism (of course, in many cases we simply haven't yet worked out what that use might be.) Many secondary metabolites are small molecules and specific to particular organisms, and this is why it is interesting to discover that chemically similar colour pigments are produced by plants, molluscs and bacteria. No one knows why Korean forest-dwelling bacteria produce blue and red pigments, just as no one knows quite why Australian sea snails also produce them.

BUGS TO FIGHT BUGS

Rough estimates suggest that since Fleming's discovery of penicillin in 1928, well over a million different natural products have been isolated from a wide range of different species. Most of these have been products with antimicrobial activities. Soil bacteria, like those from the Korean study, are a rich source of antibiotics. It's thought that they may produce them as chemical weapons to fight off other bacteria, enabling them to compete with other microorganisms for space and nutrients, and also perhaps to communicate with each other. The search for new antibiotics has become increasingly desperate with the emergence of new strains of disease-resistant microbes, such as multidrug resistant *Mycobacterium tuberculosis*.

Microorganisms themselves, therefore, may still be some of the best sources of antimicrobial drugs.

Chemists work on the principle that if they can work out how a molecule is produced in nature, they can copy the route, or even improve on it, to make their own version. A great deal of lab time is therefore devoted to mapping out the biosynthetic routes that plants, bacteria and other organisms use to make their chemicals. This is what happened in the development of the synthetic antimalarial drug, artemisinin. The natural source is sweet wormwood, but the plant can't produce the drug in the kind of quantities needed to treat the millions of people affected by malaria each year. So chemists set about characterizing the entire biosynthetic pathway, and the genes and enzymes involved. Now, they have re-engineered yeast so that it can produce the drug. The pharmaceutical company Sanofi has announced that it intends to distribute 'semi-synthetic' artemisinin as a not-for-profit venture.

Intriguingly, the biosynthetic routes that lead to the production of purple and blue dyes in nature are still not fully understood, despite the products themselves having been exploited for thousands of years. This has led some to suggest that the evolutionary coincidence that has resulted in different organisms producing very similar compounds is, in fact, no coincidence at all. As it happens, inside the snail gland from which the dye producers extract Tyrian purple is another gland and this is chock-full of bacteria. It is still only a theory, but perhaps purple bugs a bit like those found in the Korean forests have taken up residence in the glands of marine snails?

Tyrian purple

The dye Tyrian purple was used for centuries to colour the robes of royalty, and others who could afford it, before its chemical identity was finally unveiled. In 1909, German chemist Paul Friedländer got his hands on 12,000 spiny sea snails (*Bolinus brandaris*) and was able to extract 1.4 grams of the purple pigment from their hypobranchial glands. He filtered, purified and crystallized the pigment and then carried out elemental analysis, divining its chemical formula: $C_{16}H_8Br_2N_2O_2$.

The condensed idea
Nature's production line

37 Photosynthesis

Plants came up with a nifty trick when they worked out how to extract the energy from light. Photosynthesis is not only the source of all the energy that we consume in our food, it's also the source of the life-giving molecule in the air we breathe: oxygen.

Billions of years ago, our planet's atmosphere was a choking mixture of gases that, had we been around at the time, we could not breathe. There was a lot more carbon dioxide than there is today, but not much oxygen. So how did this situation change?

The answer is plants and bacteria. In fact, it's thought that the first organisms to put oxygen into the atmosphere may have been ancestors of cyanobacteria, free-floating plankton that are often referred to as blue-green algae. As the theory goes, these plankton, which produced oxygen through photosynthesis, were enslaved by plants during evolution. The cyanobacteria eventually became the chloroplasts subunits within plant cells, where the reactions of photosynthesis take place. As plants took over the planet, with the help of their cyanobacterial slaves, they pumped vast quantities of oxygen into the atmosphere. The atmosphere rapidly became one that our ancestors would evolve to breathe. Plants created an environment that humans could live in.

CHEMICAL ENERGY

Plants didn't enslave cyanobacteria for their ability to produce oxygen, though. The important product of photosynthesis, as far as plants were

TIMELINE

1754	1845	1898
Charles Bonnet notices that leaves produce bubbles when submerged in water	Julius Robert Mayer claims 'plants convert light energy into chemical energy'	'Photosynthesis' becomes an accepted term

concerned, was sugar – a molecule they could use as a fuel, a way of storing energy in chemical form. For every six molecules of oxygen made in the chloroplast, one molecule of glucose is produced.

$$6CO_2 + 6H_2O \rightarrow C_6H_{12}O_6 + 6O_2$$

carbon dioxide + water (+ light) → glucose + oxygen

This equation is actually just a summary of photosynthesis – a 'net' reaction – but what's really going on within the chloroplast is a lot more sophisticated. The green pigment, chlorophyll, which gives the leaves of plants and cyanobacteria their colour, is central to the process. It absorbs the light that kickstarts the transfer of energy from one molecule to another. The reason that plants are green is because chlorophyll absorbs only light in other parts of the visible spectrum – the green light is reflected, so it is the colour we see.

CHAIN REACTION

As light hits the chlorophyll pigments it gives them energy. This light energy is transferred from many so-called 'antenna' chlorophyll molecules to more specialized chlorophylls at the core of photosynthetic reaction centres in the chloroplasts. Electrons knocked from these specialized chlorophylls set up cascades of electron transfer, with electrons bouncing from one molecule to another like a game of pass the parcel. This redox reaction (see page 52) chain eventually leads to chemical energy being produced in the form of molecules known as NADPH and ATP, which drive the reactions that produce sugars. In the process, water is 'split' to release the oxygen that we breathe.

It's not easy – or particularly useful – to remember every single molecule involved in passing the electrons along, but the location is

NATURE HAS PUT ITSELF THE PROBLEM OF HOW TO CATCH IN FLIGHT THE LIGHT STREAMING TO THE EARTH AND TO STORE THE MOST ELUSIVE OF ALL POWERS IN RIGID FORM.

Julius Robert Mayer

1955

Melvin Calvin and colleagues map the route that carbon takes during photosynthesis

1971

First dissections of photosystems – the protein complexes involved in photosynthesis

2000

First plant genome published

Photosystems I and II

There are two types of protein complexes involved in photosynthesis in plants – one where oxygen is produced and another where the energy-carrying molecules NADPH and ATP are produced. These complexes, effectively big enzymes, are called photosystems I and II. Although it seems counterintuitive, it is easier to start by explaining photosystem II. In this photosystem, a specialized pair of chlorophyll pigments known as P680 becomes excited and kicks out an electron to become positively charged. When P680 is excited like this it is capable of accepting electrons from elsewhere, which it does by drawing them from water to release oxygen. Meanwhile, photosystem I accepts electrons passed along the chain from photosystem II, as well

as from its own light-harvesting chlorophyll molecules. The specialized pair of chlorophyll pigments in this photosystem is called P700, which also kicks out electrons to start another chain of electron transfer. Finally, these electrons flow to a protein called ferredoxin, which reduces NADP$^+$ to form the unit of chemical energy, NADPH.

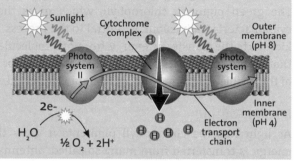

crucial. The reactions occur in assemblies of molecules called photosystems (see Photosystems, above) located in membranes within chloroplasts, the ancient cyanobacterial slaves. During the process, hydrogen ions (protons) are generated and collect on one side of the membrane. They are then pumped across the membrane by a protein – one that, conveniently, uses proton pumping to power ATP production.

CARBON FIXERS

The chemical energy (ATP and NADPH) created in chloroplasts, drives a reaction cycle that incorporates carbon dioxide from the air into sugars. They use the carbon in carbon dioxide to form the skeletons of sugar molecules. This 'carbon-fixing' process is the one that prevents our atmosphere from becoming completely clogged up with carbon dioxide. It also gives plants a sugary fuel that they can use for energy in cells, or convert into starch for storage.

You might reasonably think that plants would be quite happy about any extra carbon dioxide in the atmosphere, and that might well be the case if the only thing changing was the levels carbon dioxide, but the problem currently is that other things are changing too, such as the global temperature. With all things considered, scientists think plant growth is more likely to slow down than speed up.

BETTER THAN EVOLUTION

Plants are pretty good at extracting the energy from light, producing glucose at a rate of millions of molecules per second. But considering they've had many millions of years of evolution to hone the process, they don't actually do it that efficiently. If you compare the total amount of energy carried by the photons of light that drive photosynthesis to the amount that actually emerges in glucose, there's quite a big discrepancy. When all the energy that's lost along the way or used to drive the reactions is accounted for, the efficiency is down below 5 percent. Furthermore, that is only a maximum – most of the time the efficiency of the process is less than that.

So can humans, with less than a million years on the planet, do any better – can we extract the energy from sunlight and turn it into fuel more efficiently than plants? That's exactly what scientists are trying to do to solve our energy problems. Besides solar cells (see page 172), one idea is 'artificial photosynthesis' (see page 201) – a method of ripping water apart, like plants do, but to produce hydrogen as a fuel, or to use in reactions that make other fuels.

Energy without sunlight

In general, all the energy on planet Earth comes from the Sun and is harnessed by plants, which form the base of food chains. Plants and bacteria are autotrophs, meaning they produce their own food (sugar) and use it as an energy supply. At the bottoms of the oceans, however, where there is no light for photosynthesis, other kinds of autotrophs - chemosynthetic bacteria – extract their energy from chemicals, such as hydrogen sulfide.

The condensed idea
Plants create chemical energy using light

38 Chemical messengers

We humans have developed language as a way to communicate, but since before we could talk, our own cells have been communicating. They send messages from one part of your body to another and transmit the nerve impulses that make it possible for you to move and think. How do they do it?

The cells in your body don't work in isolation. They are constantly communicating, cooperating and coordinating their actions to help you do everything you normally do. They do this with chemicals.

Hormones control the way your body develops, your appetite, your moods and your response to danger – they might be steroid hormones (see Sex hormones, opposite), such as testosterone or oestrogen, or they might be protein hormones, such as insulin. Signalling molecules that are part of your immune system recruit cells that can also help you fight off a bout of cold or flu, but perhaps the most impressive example of how the human body uses chemical messengers is your every thought and movement, from the tiniest fluttering of your eyelids to the physical triumph of running a marathon. It's all the result of the chemical messages that are known as nerve impulses.

NERVOUS BEGINNINGS
It's not that long ago that scientists were still bickering over the nature of nerve impulses. As late as the 1920s, the most popular theory was that they were electrical, not chemical. The nerves of common laboratory animals are difficult to study because they are very delicate, so two British

TIMELINE

1877	1913	1934
Emil du Bois-Reymond ponders whether nerve impulses are electrical or chemical	Henry Dale discovers acetylcholine, the first neurotransmitter	Ethene linked to ripening of apples and pears, paving way for plant hormone research

scientists Alan Hodgkin and Andrew Huxley, decided to turn their attention to something bigger – squid nerves. Despite measuring only one millimetre in diameter, the nerves in the swimming muscles of squid were still around a hundred times thicker than those of the frogs with which they had been working. In 1939, Hodgkin and Huxley began their research on 'action potentials' – charge differences between the inside and outside of nerve cells – by carefully inserting an electrode into the nerve fibre of a squid. They found that when the nerve was firing its potential was much higher than when it was resting.

It wasn't until after the Second World War, which had thwarted their research for several years, that Hodgkin and Huxley were finally able to continue their work on action potentials. Their insights have helped us to understand that the 'electrical impulses' that travel along a nerve are the result of charged ions moving from the inside to the outside of the cell. Ion channels (see Ion channels, page 155) in the nerve cell membrane allow sodium ions to rush in as an impulse arrives, while potassium ions pour out as the impulse departs.

Sex hormones

Testosterone and oestrogen are both steroid hormones, molecules that have a wide range of effects on the body, from metabolic effects to effects on sexual development. Considering that testosterone and oestrogen are well known to play a role in the differences between male and female appearance and physiology, the structure of the two molecules is remarkably similar. Both have a four-ringed structure with only slight differences in the groups attached to one ring. Although testosterone is considered as the 'male hormone', men just make more of it, and testosterone is actually needed by women to make oestrogen – which explains why the structures are so similar. Interestingly, women's testosterone levels are highest in the morning and vary over the course of the day as well as the month, just like traditional 'female' hormones.

Testosterone

Oestrogen

How do these impulses pass from one nerve cell to the next, forming a relay chain that can pass 'messages' along? The 'message' in this case is a chain of chemical events, with each one triggering the next, like a game of Chinese whispers being played at lightning speed. Transmitting the nerve impulse to the next cell requires a molecule called a neurotransmitter to speed across the gap and stick to the membrane of the receiving cell, where it fires yet another impulse. These chemical transmission chains carry signals from our brains to the tips of our toes and everywhere in between.

> **HITLER MARCHED INTO POLAND, WAR WAS DECLARED AND I HAD TO LEAVE THE TECHNIQUE FOR EIGHT YEARS UNTIL IT WAS POSSIBLE TO RETURN TO PLYMOUTH IN 1947.**
> Alan Hodgkin on studying impulses in squid nerves

Since the discovery of neurotransmitters, beginning with acetylcholine in 1913, we have become aware of the key role that these messenger molecules play in the brain, where they are involved in the firing of 100 billion nerve cells. Treatments for mental health issues are based on the assumption that such problems have a chemical basis. In the case of depression, that assumption relates to the neurotransmitter serotonin – the antidepressant drug Prozac, launched in 1987, was thought to work by increasing serotonin levels, although that idea remains up for discussion to this day.

TALK AMONG YOURSELVES

It's not just humans and other animals that use chemical messengers, however. In any multicellular organism, the cells need ways to 'talk' to each other. Plants, for example, may not have nerves but they do produce hormones. At around the same time that physiologists were doing their groundbreaking work on nerve impulses, plant scientists were discovering that ethene is essential to fruit-ripening processes. As it turns out, ethene – the same molecule that we use to make polythene (see page 160) – doesn't only ripen fruit, it is also heavily involved in plant growth. The hormone is made by most plant cells and, like many animal hormones, transmits its signal by activating receptor molecules in cell membranes. Scientists are still unravelling the complexity of its influence on plant development and have discovered that this single hormone can switch on thousands of different genes.

Even in organisms such as bacteria that have for a long time been thought of as loners, cells must work together, and because microbes can't rely on language or behaviour to communicate, they talk using chemicals. It's only in the last decade or so that scientists have discovered that this seems to be a universal ability among bacteria. Consider, for example, what happens when you get sick. One tiny bacterium might not be able to do very much. But thousands or millions of bacteria all launching a coordinated attack is a very different prospect. How do they draw up their battle plan and muster their forces? Using chemicals – specifically, chemicals called quorum-sensing molecules. These molecules and their receptors allow bacteria that are of the same species to communicate. More broadly recognized molecules act as a kind of 'chemical Esperanto' (a universal language), allowing microbes to talk across species barriers.

Ion channels

The chemist Roderick MacKinnon was awarded the Nobel Prize in 2003 for using X-ray crystallography (see page 88) to produce 3-D structures of potassium channels. These structures helped scientists to understand the selectivity of ion channels – why one type of channel permits one type of ion (potassium) to enter while excluding another (sodium).

The myriad ways in which cells communicate using chemicals is fundamental to life. Without these signalling molecules, multicellular and unicellular organisms alike would have no way of functioning as coherent units. Each and every cell would be an island, doomed to live and die alone.

The condensed idea
Cells communicate
with chemicals

39 Petrol

Driving cars has given us the freedom to live and work as we please. Without oil and the chemical advances in petroleum refining that gave us petrol, where would we be? But petrol is also the fuel that has contributed perhaps most of all to climate change and to the pollution of our atmosphere.

On an average day in 2013, people in the USA consumed nine million barrels of gasoline (petrol). Let's say that day was 1st January. Then, the next day, on 2nd January, the USA consumed another nine million barrels and the same on 3rd January. This went on every day for 365 days, until, over the course of the year, more than three billion barrels had been consumed, just in the USA.

Most of that mind-boggling volume of petrol was burned in internal combustion engines in vehicles, which altogether travelled nearly 4.8 trillion km (three trillion miles). Now, consider that just 150 years ago, there were no cars (bar steam-powered ones), petrol-fuelled internal combustion engines had not been invented yet and the first oil well had barely been flowing five years. The rise of the motor car, fuelled by petrol, has been truly meteoric.

A THIRST FOR FUEL

Even at the beginning of the 20th century, there were only eight thousand registered cars in the whole of the USA and they were all trundling along at less than 32 km/h (20 miles an hour). But by this time, the oil rush had begun

TIMELINE

1854	1859	1880	1900
Pennsylvania Rock Oil Company formed, produces oil by digging and trenching	Drilling of first oil well	First petrol-powered internal combustion engine	Number of registered cars in the USA tops 8,000

and oil tycoons like Edward Doheny – who is said to have inspired Daniel Day Lewis's character in the film *There Will Be Blood* – were making their millions. Doheny's Pan American Petroleum & Transport Company drilled Los Angeles' first free-flowing oil well in 1892. By 1897, there were 500 more.

The demand for petrol was growing faster than chemists' knowledge of petroleum. In 1923, writing in *Industrial and Engineering Chemistry*, Carl Johns of the Standard Oil Company of New Jersey bemoaned the lack of chemical research in the area. Meanwhile, Hollywood celebrities and oil-made millionaires, including the Dohenys, were driving around expensive cars. Edward's son Ned had bought his wife a car designed by Earl Automobile Works. It was battleship grey with red leather upholstery and Tiffany lamps. Earl Automobile's chief designer Harley Earl eventually moved to General Motors where he took charge of its Art & Colour department and went on to style Cadillacs, Buicks, Pontiacs and Chevrolets.

> I HAD FOUND GOLD AND I HAD FOUND SILVER … BUT THIS UGLY-LOOKING SUBSTANCE I FELT WAS THE KEY TO SOMETHING MORE VALUABLE THAN … THESE METALS.
>
> Edward Doheny

BURNING AMBITION

Thanks to increasing demand for cars and Henry Ford's resolve to satisfy it through his assembly-line scheme for mass production, petrol stations began to pop up all over the road network. Advances in petroleum refinery processes, including cracking (see page 60), soon meant that petrol producers were able to obtain high-quality blends of petrol that burned more smoothly.

The mixture that fills up your car's fuel tank today contains hundreds of different chemicals, including a blend of hydrocarbons, as well as additives such as anti-knocking, anti-rusting and anti-icing agents. 'Hydrocarbons' cover a huge array of straight-chain, branched, cyclic (ring-structured) and aromatic (see Benzene, left) compounds. The

1913	1993	2000	2014
The Ford Motor Company begins the first moving automobile assembly line	Euro 1 emissions standards applying to passenger cars come into force	Number of registered motorway vehicles in the USA reaches 226,000,000	Euro 6 emissions standards come into force

Benzene

Benzene is a ring-structured hydrocarbon that is produced during the petroleum-refining process, and is present naturally in crude oil. It is an important chemical industrially, in the production of plastics and drugs. The benzene ring of six carbon atoms is stable, and is also found in a wide variety of natural and synthetic compounds called aromatic hydrocarbons. Paracetamol and aspirin are examples of aromatic benzene derivatives, as are the sweet-smelling compounds in cinnamon bark and vanilla. Benzene itself is carcinogenic and levels in petrol are strictly controlled to prevent dangerous atmospheric emissions. Improvements to catalytic converters have played an important role in reducing benzene emissions.

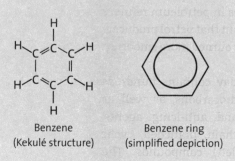

| Benzene | Benzene ring |
| (Kekulé structure) | (simplified depiction) |

chemical identity of the components depends partly on where the oil originally came from. Crude oil from different regions of the world, with different properties, is often blended together.

In the combustion engine of a car, petrol burns in air, which provides the oxygen required for combustion, to produce carbon dioxide and water. For example:

$$C_7H_{12} + 11O_2 \rightarrow 7CO_2 + 8H_2O$$

heptane + oxygen --> carbon dioxide + water

This is an example of a redox reaction (see page 52), because carbon atoms in heptane are oxidized, while oxygen is reduced.

POLLUTION PROBLEMS

Until a few decades ago, the anti-knocking effects of tetraethyllead in leaded petrol prevented the fuel exploding before it reached the working part of the engine, allowing more efficient combustion. But the addition of tetraethyllead also meant that car exhausts pumped toxic lead bromide into the atmosphere – a result of tetraethyllead reacting with another additive, 1,2-dibromoethane, which was intended to stop lead clogging up the engine. Leaded petrol was phased out starting in the 1970s, while petrol producers sought new ways to make smooth-burning, high-octane fuels (see Octane numbers, opposite) that would burn for more miles per gallon.

So that was one problem in hand, but as the car industry boomed during the 20th century, levels of carbon dioxide in the atmosphere soared. Levels of other pollutants rose too, because the energy supplied by the car engine causes other components of air to react. Nitrogen reacts with oxygen to produce nitrogen oxides (NO_x), which form smog and cause lung disease. Around half of all NO_x emissions are thought to be due to road transport.

CHEMICAL SOLUTIONS

Cutting vehicle emissions has become a priority for car manufacturers, as increasingly more stringent limits are imposed. As car manufacturers entertain the possibilities for electric and hybrid vehicles, solutions are still needed for ordinary petrol- (and diesel-) fuelled cars. The three billion barrels of petrol that are burned every year, just in the USA, are enough to fill over two hundred thousand Olympic-size swimming pools. All that petrol adds up to more than a gallon per day (3.8 litres/day) for every US citizen. Catalysts for catalytic convertors, NO_x traps and other vehicle emissions reduction technologies are now active areas of research for chemists.

> ### Octane numbers
>
> The octane number of a petrol mixture, or a particular component of petrol, is a measure of how smoothly and efficiently it burns. Octane numbers are measured relative to 2,2,4-trimethylpentane (or 'iso-octane'), which is considered high-octane at 100, and to heptane, which has an octane number of 0. Those components of petrol that have low octane numbers are more likely to cause engine knocking.

Chemical advances have allowed the production of more efficient fuels, which in turn have allowed the car-driving masses to travel further, more affordably. Now chemistry is having to deal with the consequences: an atmosphere choked with exhaust gases and the dwindling of resources with which to fuel our everyday commutes.

The condensed idea
Fuel that changed the world

40 Plastics

Whatever did do we do before the invention of plastic? What did we carry our shopping home in? What did we eat crisps out of? What was everything made of? It's odd to think that time was not so long ago.

When crisps (potato chips) were first mass produced they were sold in tins, wax-coated paper packets or, sometimes, large bins out of which they were scooped, like pick 'n' mix. Today, buying crisps is more convenient and hygienic – they're sold in plastic packets, just like most of the other food we buy.

America's first potato chip company was founded in 1908, the year after the first fully synthetic plastic, Bakelite, was invented. Bakelite is an amber-coloured resin made by reacting together two organic compounds, phenol and formaldehyde. Initially, at least, the plastic was used in all sorts of products, from radios to snooker balls. The Bakelite Museum in Somerset, England, even boasts a Bakelite coffin. It's thermosetting, meaning that once it sets, it can't be reshaped by heating.

THE MATERIAL OF A THOUSAND USES.

Bakelite company slogan

Within a few decades, a whole range of other plastics, including various remouldable (thermoplastic) plastics, had become available. For a while, it was thought that these new, durable materials were made of close aggregations of short-chain molecules, but during the 1920s, the German chemist Hermann Staudinger came up with the concept of 'macromolecules' and proposed that plastics were actually made of long polymer chains (see page 16).

TIMELINE

3500BC	1900	1907	1922
'Natural plastic', tortoiseshell, used by Egyptians to make combs and bracelets	Recognition of polymers	The Age of Plastic begins with Bakelite, the first fully synthetic plastic	Hermann Staudinger proposes plastics are made of long-chain molecules

THE AGE OF PLASTIC

In the 1950s, the polythene bag – the most ubiquitous product of the plastic era – arrived on the scene. The Age of Plastic was in full swing. Soon, crisps and other food items were being sold in plastic packets, meaning an entire week's shopping could be brought home festooned in plastic.

The process for making polythene was revealed in an accidental discovery by British scientists at ICI in 1931. This involved heating ethylene (also known as ethene) gas at high pressure to produce what is alternatively called polyethylene, a polymer of ethylene. Ethene is a product of the chemical cracking of crude oil (see page 60), so most polythene has its origins in the petroleum industry. However, ethylene – and, therefore, polythene – can also be made using renewable resources, for instance, via a chemical conversion of alcohol produced from plants like sugar cane.

The majority of polythene bags are made from low-density polythene (LDPE), produced at high pressure, as in the ICI process. The polymer chains in LDPE are straight, whereas high-density polythene (HDPE), which is produced at low pressure, contains branched molecules that form a stiffer material.

Natural plastics

Natural materials that behave a bit like plastics are sometimes referred to as natural plastics. For example, animal horn and tortoiseshell (from the shells of marine turtles) can, like plastics, be heated up and moulded into a desired shape. In fact, these materials are not really what we would think of as plastics. They are composed mostly of a protein called keratin – the same protein found in our hair and nails. Like a plastic, however, keratin is a polymer containing many repeated units. Because it is now illegal to trade in many of these materials, the tortoiseshell that was once used to make combs and other hair ornaments has been replaced almost entirely by synthetic plastics. The first imitation tortoiseshell was celluloid, a semi-synthetic material invented in 1870 that also functioned as a useful substitute for the ivory used to make snooker balls. It tended to catch fire very readily – so much so, in fact, that it was soon superseded by the slightly less flammable 'safety celluloid'. Today, newer plastics such as polyester are used as tortoiseshell substitutes.

1931	1937	1940	1950s	2009
Accidental discovery of polyethylene (polythene)	Commercial production of polystyrene	PVC goes into production in the UK	Polythene bags	Boeing's 787 aeroplane is 50% plastic

THE DOWNFALLS OF DURABILITY

To begin with, the environmental implications of escalating plastics production weren't given much thought. Plastics were, after all, chemically inert – they lasted a long time and didn't seem to react with anything in the environment. However, this attitude led to soaring volumes of plastic waste entering landfill sites as well as the oceans. In the North Pacific Ocean, there is a rotating 'trash vortex' of untold size, composed mostly of plastic. It's thought that each square kilometre of water in this area contains around three quarters of a million pieces of microplastics, little particles of plastic that fish can mistake for plankton.

Many plastics are non-biodegradable, breaking up over time into smaller pieces or microplastics. On land, these microplastics can block the guts of birds and mammals. Polythene is about the least biodegradable plastic there is. 'Green polythene' made from sugar cane is just the same (see Bioplastics, opposite). However, views on biodegradation among chemists and microbiologists are now shifting slightly.

PLASTIC-EATING MICROBES

The reason that polythene hangs around in the environment is that it isn't broken down by microbes. This is because its structure, being composed entirely of chains of carbon and hydrogen, doesn't contain any of the chemical groups that they like to utilize. Microbes attach to oxygen-containing groups like carbonyl (C=O). Oxidation, using heat and catalysts – or even sunlight via photo-oxidation – is one way of converting polythene into a form that bugs are more easily able to digest. But another option is simply to look for specific bugs that aren't so bothered about the oxidized bits.

Microbiologists have now discovered bacteria and fungi that make enzymes capable of degrading or 'eating' plastics. Some can actually grow in films on the surface of polythene, using it as a source of carbon for metabolic reactions. In 2013, Indian scientists reported that they had found three different species of marine bacteria in the Arabian Sea that could degrade polythene

Bioplastics

The term bioplastics is confusing. Sometimes it means plastics made from renewable materials, such as plant cellulose – more accurately called bio-based plastics – and other times it means biodegradable plastics. Poly(lactic acid) (PLA) is made from plant material and is biodegradable. However, not all bio-based plastics are biodegradable. Polyethylene can be made from plant materials, but is extremely resistant to biodegradation.

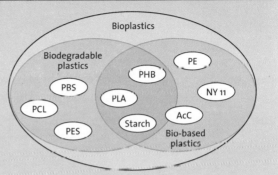

without it first being oxidized. The best was a subspecies of *Bacillus subtilis*, a microorganism commonly found in soil and the human gut. Meanwhile, the Indian nation alone continues to consume 12 million tonnes of plastic products every year and tens of thousands of tonnes of waste each day.

The reason crisp packets often can't be recycled, however, is that they contain a layer of metal, for 'added freshness' which keeps oxygen out. Short of shredding them yourself and 'upcycling' them into designer garments, you have to send them to landfill. However, the plastic most often used in crisp packets is polypropylene and, in 1993, Italian chemists found they could get bacteria to grow on polypropylene by adding sodium lactate and glucose. Theoretically, perhaps we could get microbes to eat our crisp packets, as well as other plastic waste. But the biggest impact could be made on waste simply by reducing the amount of plastic packaging we use.

The condensed idea
Multipurpose polymers causing a pollution problem

41 CFCs

For years, CFCs were considered safe alternatives to the poisonous gases originally used in refrigerators. There was only one problem: they destroyed the ozone layer. Before this problem was fully recognized and accepted, however, the hole in the ozone layer had grown to the size of a continent. The commercial use of CFCs was eventually banned in 1987.

The refrigerator has been in our homes for less than a century, but it has become so engrained in daily life that we now take it for granted. We can drink a glass of cold milk whenever we feel like it and the gently humming box in the corner of the kitchen has inspired culinary masterpieces, such as the chocolate fridge cake. In 2012, the Royal Society decreed the refrigerator the most important invention in the history of food.

While it's certainly a relief not to be restocking your pantry every other day, there's always the chance that you might discover something unsavoury lurking at the back of your refrigerator. What if instead of a few rotting leaves of lettuce it was a continent-sized hole in the ozone layer?

We now know that the gases responsible for the hole in the ozone layer were CFCs – refrigerants developed to replace the poisonous gases being used in fridges in the early part of the 20th century. These chlorine-containing compounds break down in sunlight to release damaging chlorine-free

TIMELINE

1748	1844	1928	1939
First demonstration of refrigeration	John Gorrie builds an 'ice maker'	CFCs developed for refrigerators	First fridge-freezer in USA

How did CFCs destroy the ozone layer?

In sunlight, CFCs break down to release chlorine radicals – free chlorine atoms that are very reactive due to their unpaired electrons or 'dangling bonds'. The chlorine radicals kickstart a chain reaction that pulls oxygen atoms away from ozone (O_3) molecules. They temporarily team up with oxygen to form compounds of chlorine and oxygen but are then recycled to produce more chlorine radicals, which destroy more ozone molecules. Similar reactions occur with bromine. During the Antarctic winter, there is very little or no sunlight, so it is only when spring arrives and daylight returns that the reactions occur. The rest

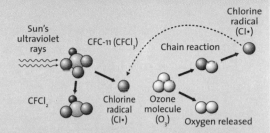

of the year, the chlorine from CFCs remains locked up in stable compounds in icy clouds. Ozone can also be broken down naturally by sunlight, but it usually reforms at about the same rate. When chlorine radicals are present, however, they tip the balance in favour of ozone destruction.

radicals into the atmosphere (see How did CFCs destroy the ozone layer?, above). Before CFCs, manufacturers of refrigerators were using methyl chloride, ammonia and sulfur dioxide, all of which are very dangerous if inhaled in an enclosed space. A refrigerant leak could be lethal.

COOL SOLUTION

Many accounts cite a deadly explosion involving methyl chloride at a hospital in Cleveland, Ohio, in 1929 as the motivation for developing non-toxic refrigerant gases. In actual fact, the 120 people killed in that disaster seem to have died due to inhalation of carbon monoxide, along with nitrogen oxides produced when some X-ray film caught fire, rather than methyl chloride. But in any case, the chemical industry was already well aware of the hazards of using poisonous gases as refrigerants and was working on a solution.

1974

Discovery of mechanism of ozone layer depletion by CFCs

1985

Hole found in ozone layer over Antarctica

1987

Agreement under Montreal Protocol to reduce use of ozone-depleting chemicals

The year before the Cleveland accident, Thomas Midgley, Jr., a researcher at General Motors, had made a non-toxic, halogen-containing compound called dichlorodifluoromethane (CCl_2F_2), an ungainly name that was shortened to 'Freon'. This was the first CFC, although it wasn't reported publicly until 1930. Midgley's boss, Charles Kettering, was looking for a new refrigerant that would 'not take fire and which would be free of harmful effects on people'. Retrospectively, it might be considered something of a bad omen that it was Midgley, fresh from the discovery of tetraethyl lead – the anti-knocking agent in leaded petrol – who was assigned the task.

> **LIVING ON $6 A DAY MEANS YOU HAVE A REFRIGERATOR, A TV, A CELL PHONE, YOUR CHILDREN CAN GO TO SCHOOL.**
> Bill Gates

In 1947, three years after Midgley's death by probable suicide, Kettering wrote that Freon had just the properties that were required. It did not catch fire and was 'altogether without harmful effects on man and animals'. This was true in one sense: it caused no direct harm when people or animals were exposed to it. Kettering noted that none of the lab animals used in testing showed any signs of illness when they were allowed to breathe in the gas. Midgley had even demonstrated how safe CFCs were by breathing them in himself, taking a big gulp of the gas during a presentation. So it came to be that CFCs were adopted as the new refrigerants. Midgley, having met an early demise, did not survive to understand the impact of his research.

PLUGGING THE HOLE

In 1974, at about the time that fridge-freezers were being stuffed with Black Forest gateaux and Arctic rolls, the first evidence of the effects of CFCs emerged in a paper published by Sherry Rowland and Mario Molina, two chemists at the University of California. The paper claimed that the ozone layer – which filters out the most harmful parts of UV radiation coming from the Sun – could be depleted by half before the middle of the 21st century unless CFCs were banned.

Unsurprisingly, these claims were met with consternation on the part of the chemical companies that were making money from the refrigerants. At this point, there was still no proof that CFCs had done any actual damage to the ozone layer – Rowland and Molina had only described a mechanism. Many people were still sceptical about the idea and argued that the economic consequences of banning CFCs would be grave.

It was another decade before conclusive proof for the hole in the ozone layer was provided. The British Antarctic Survey had been monitoring ozone in the atmosphere above Antarctica since the late 1950s and by 1985 scientists had enough data to know that levels were dropping. Satellite data showed that the hole extended over the entire continent of Antarctica. Just a couple of years later, countries around the world ratified the Montreal Protocol on Substances that Deplete the Ozone Layer, which set out a timetable for the phasing out of CFCs.

What about now?

The hole in the ozone layer increased dramatically in size between the end of the 1970s and the early 1990s. Since then, with the signing of the Montreal Protocol, its average size has plateaued and finally started to decrease. The hole was at its largest in September 2006, at about 27 million square kilometres. Since ozone-depleting chemicals in the atmosphere are long-lived, according to NASA scientists it may take until 2065 before the hole shrinks to the size it was in the 1980s.

What's lurking in the back of your refrigerator these days then? Some manufacturers have replaced CFCs with HFCs (hydrofluorocarbons). Since it's the chlorine that does the damage, hydrofluorocarbons are a common substitute. However, in 2012, Mario Molina was one of the authors of a paper highlighting another problem: HFCs may not damage the ozone layer, but some of them are more than a thousand times more potent as greenhouse gases than carbon dioxide. In July 2014, for the fifth year in a row, parties to the Montreal Protocol discussed its extension to HFCs.

The condensed idea
A cautionary tale about chemicals

42 Composites

Why use one material when two are better? Combining different materials can produce hybrid materials with extraordinary properties, such as the ability to withstand temperatures up to thousands of degrees or absorb the impact of a bullet. Advanced composites protect astronauts, soldiers, the police force and even your delicate smartphone.

On 7th October 1968, the first manned Apollo spacecraft launched from Cape Kennedy Air Force Station in Florida, beginning a tense, 11-day flight that would test relations between crew and mission control. The previous year, three crew members had died in the launch of the only other manned Apollo mission. The remaining Apollo missions proved successful, however, not only because they put humans on the Moon for the first time, but because they delivered their crews safely back to Earth.

One key safety feature of the Apollo command module was its heat shield. When an explosion crippled Apollo 13, forcing the crew to make their way home under limited power, their fate depended on the heat shield. Before re-entry into Earth's atmosphere no one was sure whether the heat shield was still intact. Without the protection it afforded, Jim Lovell, Jack Swigert and Fred Haise would have fried.

IN THE MATRIX
The heat shields on the Apollo mission command modules were made from composite materials that are said to be 'ablative' – they burn up slowly, while protecting the spacecraft from damage. The particular composite they

TIMELINE

1879	1958	1964	1968
Thomas Edison bakes cotton to make carbon fibres	Roger Bacon demonstrates first high-performance carbon fibres	Stephanie Kwolek develops aramid fibres	Apollo command module uses composites in manned space flight

Kevlar®

There are various types or grades of Kevlar fibres, some stronger than others. Mostly, we hear about those that are used as reinforcement in lightweight bulletproof materials, but the fibres are also used in the hulls of boats, wind turbines and even the cases of some smartphones. Chemically, the polymer chains in Kevlar are not dissimilar to nylon – both contain a repeating amide group, highlighted in the chemical structure below. Stephanie Kwolek was working with nylon when she invented Kevlar at DuPont. In nylon, however, the chains get twisted so they can't form such stable sheets. Each amide group in a Kevlar polymer chain can form two strong hydrogen bonds connecting it to two other chains. Repeated along the length of each chain, this creates a regular arrangement of great strength.

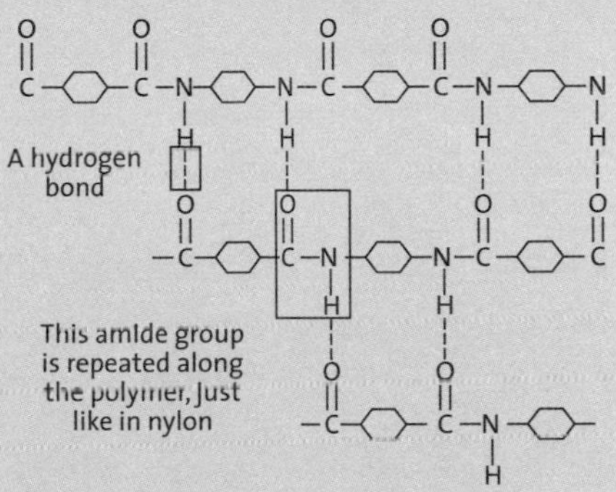

A hydrogen bond

This amide group is repeated along the polymer, just like in nylon

Structure of Kevlar

One downside, however, is that this structure also makes the material stiff, so a bullet-proof vest may save your life but it probably won't be very comfortable.

employed was called Avcoat and while it hasn't been used in space flight since the Apollo missions, NASA has announced plans to use it on the heat shield of Orion, which will be the next manned spacecraft to visit the Moon.

Like other composites, Avcoat's special properties – such as being able to resist temperatures up to thousands of degrees – are a result of its combination of materials. Together, the different materials form a new supermaterial that's greater than the sum of its parts. Many composites are made up of two main components. One is the 'matrix', which is often a resin that acts as a binder for the other component. This second component is usually a

1969
F-4 jet plane gets boron-epoxy rudders

1971
Kevlar® aramid fibres marketed by DuPont

2015
Orion spacecraft due to launch with Avcoat composite heat shield

fibre or fragment that reinforces the matrix, giving it strength and structure. Avcoat is made of silica fibres embedded in a resin that is then formed into a fibreglass honeycomb structure. For the Apollo command modules, there were over 300,000 holes in the honeycomb and the filling process was carried out by hand.

COMMON COMPOSITES

You might think you don't know any other materials like Avcoat. But composites are not only used in space flight and are more common than you might think. Concrete is a good example of a composite material. It's formed from a combination of sand, gravel and cement. There are also natural composites like bone, which is made from a mineral called hydroxyapatite and the protein collagen. Materials scientists are trying to mimic the structure of bone to develop new composites, such as advanced, nanostructured materials with potential medical applications.

> **I THOUGHT, THERE'S SOMETHING DIFFERENT ABOUT THIS. THIS MAY BE VERY USEFUL.**
> Stephanie Kwolek
> on inventing Kevlar®

Perhaps the most widely recognized composites are carbon fibre and Kevlar. The name carbon fibre refers to the stiff carbon filaments that give golf clubs, Formula 1 racing cars and prosthetic limbs their strength. Discovered in the 1950s by Roger Bacon, they formed the first high-performance composite materials. (Concrete had begun to enter widespread use a century earlier.) Bacon called his filaments carbon 'whiskers' and showed they were 10 to 20 times stronger than steel. Usually, when we say carbon fibre, we are referring to carbon-fibre-reinforced polymer, a composite that is formed when the whiskers are embedded in a resin like epoxy, or some other binding material.

A few years later, aramids were discovered by chemist Stephanie Kwolek at the American company DuPont, which acquired patents and marketed her material as Kevlar® (see Kevlar®, page 169) in the 1970s. Kwolek originally discovered the bulletproof fibres while working on materials for tyres – she found she could make a fibre that was tougher than nylon and would not break when it was spun. The reason for Kevlar's strength is related to its very regular, flawless chemical structure, which in turn promotes regular hydrogen bonding (see page 20) between polymer chains.

TAKING FLIGHT

High-performance composites like carbon fibre are not only found in spacecraft. A a modern aeroplane may be a patchwork of different composites. The main body of Boeing's 787 Dreamliner is 50 percent advanced composites – mostly carbon-fibre-reinforced plastic. These lightweight materials add up to a 20 percent weight saving overall, compared to a more conventional aluminium aircraft.

Weight savings also offer an advantage on the ground and in 2013, engineers at Lynchburg-, in Virginia, USA, based company Edison2 unveiled the fourth edition of its VLC – very light car. The VLC 4.0 weighs just 635 kilograms – less than a Formula 1 racing car and about half the weight of an ordinary family-sized car – although it looks more like a very tiny, white plane. Like the Dreamliner, it combines steel, aluminium and carbon fibre.

After a decade in development, NASA's Orion spacecraft is almost ready for its first unmanned test flights. The safety of later manned flights – like the earlier Apollo spacecraft – will depend on the command module's Avcoat heat shield. At five metres in diameter, Orion's heat shield is purported to be the largest ever made. The manufacturing process has had to be 'recovered'; some of the original ingredients are not even available today. Yet Avcoat is still considered the best material for the job some 50 years on.

Self-healing materials

Imagine a plane wing that could heal its own cracks. One much talked about application of composites is in self-healing materials. Researchers at the University of Illinois at Urbana-Champaign, USA, have been working on fibre-reinforced composite materials that contain channels filled with healing agents – so if a material gets damaged, the channels release a resin and a hardener that, when combined, seal it back together. In 2014, they reported a system that could repeatedly self-heal in this way.

The condensed idea
Materials greater than the sum of their parts

43 Solar cells

Most modern solar panels are made from silicon, but scientists are working to try to change that. They want something cheaper and, well, more 'see-through', perhaps based on a composite material. Even better would be something that could be applied as a spray-on coating, so you could put it on any glass surface – imagine if you could run your radiators off your windows!

It's The Future. You're buying a brand new house and you're being asked to make all sorts of difficult decisions. Which tiles do you want in the bathroom? Standard taps or fancy ones? What colour carpets? There are also options for the windows: you choose double-glazed, but you wonder about solar. The property developer tells you that if you choose solar, someone employed by the window supplier will spray a completely transparent, light-absorbing substance onto the glass panes you've ordered. Your solar windows will generate electricity that can be fed back into the National Grid and used to pay for up to half of your heating bill. And they won't look any different to normal windows.

That's the dream anyway. Back in the present, we're still dealing with difficult issues like efficiency – how to extract the maximum amount of energy from sunlight – and the cost of making these materials. But it's not so far-fetched to imagine spray-painting windows, and other household surfaces, with materials that harvest sunlight. Much of the work has already been done in the lab, at least.

TIMELINE

1839	1839	1954	1958
Photovoltaic effect observed by Edmond Becquerel	'PN barrier' observed by Edmond Becquerel	Bell Labs researchers invent the silicon solar cell	First satellite (Explorer VI) with a photovoltaic array is launched

Dye-based solar cells

In photosynthesis, light energy is extracted by chlorophyll, a natural pigment that gets excited by sunlight and passes this excitement along as electrons, through a series of chemical reactions, to create chemical energy (see page 148). Dye-sensitized solar cells, invented by the Swiss chemist Michael Grätzel in 1991, do a similar sort of thing using the pigment molecules of dyes. Dye-sensitized refers to the fact that it is the dye that makes the cell sensitive to light. The dye is coated onto a semiconductor within the solar cell – the two are chemically bonded – and when the light strikes the dye, some of its electrons become excited and 'jump' into the semiconductor layer, which conducts them into an electric current. Scientists have tested dyes that are porphyrins, like the chlorophyll pigments in plants. The best photosensitive dyes are considered

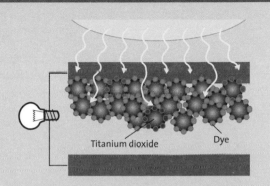

Titanium dioxide   Dye

to be those containing transition metals like ruthenium, though ruthenium is a rare metal, so it doesn't exactly lend itself to the sustainable manufacture of solar panels. Efficiency is also generally low. In 2013, however, Grätzel's own team at the Swiss Federal Institute of Technology used perovskite materials to increase the energy-extracting efficiency of its dye-sensitized cells to 15 percent.

STARTING WITH SILICON

Today, most of the solar panels you see on buildings or in solar parks are made of silicon – unsurprising, as due to the ubiquity of silicon in computer chips, we already know a lot about the chemistry and electronic properties of the material. The first silicon solar cell, or 'solar battery' as it was dubbed by its creators, was made at Bell Labs, the semiconductor firm that developed the transistor and techniques for patterning silicon that would become

1960
Silicon Sensors starts producing silicon solar cells

1982
First megawatt-scale solar power station

1991
Michael Grätzel and Brian O'Regan report the first dye-based solar cells

2009
First reports of perovskites in solar cells

crucial to the manufacture of silicon chips (see page 96). This solar battery was announced in 1954 and could convert the energy from sunlight with an efficiency of about 6 percent. It was soon powering space satellites.

Research into the photovoltaic effect, which was first discovered in 1839 by the French physicist Alexandre-Edmond Becquerel, is deeply rooted in the history of Bell Labs and the chemist Russell Ohl. In 1939, Ohl was searching for materials capable of detecting short-wave radio signals. While taking some electrical measurements in silicon, he turned on a cooling fan in the lab. It was positioned between the window and his silicon cylinders. Oddly, the voltage spikes he measured seemed to coincide with the rotation of the fan blades letting light through. After some head-scratching, Ohl and his colleagues realized that silicon conducts a current when exposed to light.

Today, while the very best silicon photovoltaic technology is pushing towards 20 percent efficiency, it is still pretty expensive and there's no chance of slapping it on your windows. But the dream of efficient 'building-integrated photovoltaics' has become more realistic since the development of organic solar cells, which, like plants, use organic molecules (see page 148) to capture the energy in sunlight. These organic solar cells can be made in large, thin, flexible films that can be rolled up, bent or wrapped around a curved surface. The only problem is that they're currently not as efficient as inorganic silicon solar cells.

Perovskites

Perovskites are hybrid organic/inorganic materials containing halogens like bromine and iodine, and metals. One of the perovskites that has seen the most success so far in solar cells has the chemical formula $CH_3NH_3PbI_3$ – it also contains lead. This is a problem because lead is toxic and environmental legislation aimed at reducing lead use in products like paint has been in place for decades. On the other hand, researchers recently showed they could recycle the lead from old batteries to make perovskite solar cells.

GOING ORGANIC

The basic architecture of an organic solar cell is a sandwich, where the two slices of bread are electrode layers and the filling is made up of layers of organic materials that are activated by sunlight. UV light excites electrons in the material, transporting them to the electrodes and generating a current. Enhancing the materials used in the inner or outer layers of the sandwich could create a more efficient solar cell. Graphene (see page 184), for instance, has been tested as an alternative to commonly used indium tin oxide electrodes and can work just as well, according to a US study published in 2010. Both are transparent, but carbon-based graphene would be preferable because indium tin oxide is in limited supply.

The chemical company BASF recently joined forces with Daimler, a division of Jaguar, to make organic, transparent, light-harvesting solar arrays for the roof of its new electric car, the Smart Forvision. Unfortunately, the roof doesn't absorb enough energy to fuel the car, but it can just about power the cooling system. It's still the overall efficiency of organic solar cells that plagues their development and practical use. They're not yet able to go much beyond 12 percent. Plus, while a silicon solar panel might last up to 25 years, an organic equivalent would struggle to reach half that age. On the other hand, they can be made in all almost any colour, and they're bendy. So if you're interested in purple-tinted, bendy, solar-powered devices that you can throw away after a couple of years, organics are probably your thing.

> I'D PUT MY MONEY ON THE SUN AND SOLAR ENERGY, WHAT A SOURCE OF POWER! I HOPE WE DON'T HAVE TO WAIT UNTIL OIL AND COAL RUN OUT, BEFORE WE TACKLE THAT.
>
> Thomas Edison

SPRAY-ON SOLAR

While research on organic materials has focused on improving their efficiency and lifetime, a new material has sprung onto the scene. Perovskites (see Perovskites, opposite) were rated among the top ten scientific breakthroughs of 2013 by the internationally renowned journal *Science*. These hybrid organic/inorganic materials have quickly achieved surprising levels of efficiency of 16 percent, and are shooting for 50 percent, apparently. They're easy to make and what's more, techniques for spray-coating them onto surfaces are already in development. Perhaps the windows of 'The Future' are not so far away. Though paying for half of your heating bill is definitely a tall order.

The condensed idea
Materials that make electricity with sunlight

44 Drugs

How do chemists go about making a drug? Where does the idea come from, and how is it turned into a functional chemical compound or mix? Many of the products of the pharmaceutical industry are based on natural chemicals, while others are the 'hits' generated by screening thousands or millions of different compounds for those that do the job that's required.

There are lots of different kinds of drugs. There are the kind of drugs that a doctor prescribes. There are the kind of drugs dished out by shady characters in back alleys. There are drugs that kill you. Drugs that cure you. There are uppers. There are downers. There are drugs that come from fungi, venomous snails, poppies and willow bark. There are completely synthetic drugs designed and made by chemists. And then there are the one-of-a-kind drugs based on compounds found in sea sponges, that come in half a million different chemical forms, take 62 separate chemical steps to make and are used to treat advanced breast cancer.

ALL AT SEA

In the early 1980s, Japanese researchers at Meijo and Shizuoka Universities were collecting samples of sponges from the Miura Peninsula south of Tokyo. Sponges are aquatic animals – colonies containing hundreds or thousands of individuals that look more like plants or mushrooms. One particular animal – a black sponge that the researchers had collected 600 kilograms of to experiment with – produced a compound that sparked their interest. In 1986, they announced in a chemistry journal that this compound 'exhibited remarkable ... anti-tumour activity'.

TIMELINE

1806	1928	1942	1963
Morphine isolated from opium poppies	Discovery of penicillin	Relative of chemical weapon, mustard gas, used as first cancer chemotherapy	Launch of benzodiazepine (Valium)

Viagra

Sildenafil, better known as Viagra, is a drug described as a 'phosphodiesterase type 5 inhibitor' – it stops an enzyme called phosphodiesterase type 5 (PDE5) from doing its job. By the 1980s, Pfizer scientists already knew that PDE5 was responsible for breaking down a chemical that causes muscles in blood vessels to relax. Viagra works by stopping PDE5 from degrading this chemical, allowing blood to rush into the relaxed blood vessels. The Pfizer team were originally working on a treatment for heart disease. In 1992, they began testing sildenafil on heart patients. Two things quickly became apparent: first, the drug wasn't particularly useful for treating blood pressure or angina; and second, it had some unusual side effects in male patients.

Viagra molecule

In the past, there would have been precious few options for harnessing the power of such a compound, besides harvesting even more sea sponges. And that is, initially at least, what people tried to do. After it emerged that another, more common, deep-sea sponge produced the same cancer-beating chemical, the National Cancer Institute (NCI) in the USA and New Zealand's National Institute of Water and Atmospheric Research funded a half-a-million-dollar project to lift a tonne of the animal from the seabed off the coast of New Zealand. This got them less than half a gram of the compound they were after – halichondrin B.

Even worse, halichondrin B seemed virtually impossible to copy using synthetic strategies. It was a large, complex molecule with billions of different forms – these stereoisomers (see page 137), where the same atoms are connected to each other, but with some of the chemical groups in different orientations.

1972	**1987**	**1998**	**2006**
Discovery of fluoxetine (Prozac)	First statin, lovastatin, available for prescription	Viagra launched	Sales of Pfizer's cholesterol-lowering drug Lipitor peak at $13.7 billion

IN THE LIBRARY

By the time the 1990s arrived, chemists had hit upon a new strategy for making drugs. Instead of relying on natural biosynthesis (see page 144) or long-winded chemical synthesis (see page 64) of a specific molecule, they were generating whole 'libraries' of different molecules and screening all of them for interesting activities. This method can be useful if you want, say, a molecule to target a specific receptor on a cell (see Easy target?, below). Using a chemical library, you can run the same test on lots of different molecules and produce a list of the ones that target that receptor. Now you have your shortlist, you can study each one more carefully.

Easy target?

Most of the bestselling drugs are chemicals that target cell-surface receptors, like, for instance, the GPCRs – G-coupled protein receptors. The GPCRs are a huge group of receptors that sit in the membranes of cells, where they pass on chemical messages. More than a third of all prescription drugs – including Zantac, for indigestion and the schizophrenia drug Zyprexa – target GPCRs. That's why drug developers continue to screen thousands of potential drugs at a time, looking for any that might affect GPCRs.

Meanwhile, a chemical route for synthesizing halichondrin B had eventually been published, but it was tedious and still didn't make enough of the compound. A Japanese company called Eisai Pharmaceuticals started churning out compounds a bit like halichondrin B, but less complex, to see if they could find one that worked just as well. They were intended as analogues, in that their mode of action should be the same, even if their structures were different. Eisai's scientists knew from the NCI's work that the original compound acted on tubulin, a protein that holds together the structure of cells and that is needed for cancer growth. Any effective analogue would have to target the same protein.

While their method may have been a been a bit old-fashioned, it worked. They found eribulin, a product that is now licensed for treatment of advanced breast cancer, even if it does have over half a million possible stereoisomers and take 62 steps to make. Taking your inspiration from nature is still the best way to be successful in the drug business because nature has already done most of the work. About 64 percent of all new drugs licensed between 1981 and 2010 had some kind of natural inspiration. Most are either extracted from living organisms, modelled on or modified from chemicals made by living organisms, or designed specifically to interact with specific molecules in living organisms. Sometimes it just takes a bit (or a lot) of clever chemistry to put that inspiration to good use.

DRUGS BY DESIGN

Even so, there are plenty of successful drugs that come from elsewhere. Take Viagra (see Viagra, page 177), a failed blood-pressure drug that became the fastest-selling drug of all time. But if you need somewhere to start looking, the obvious place is often the natural molecules that are responsible for disease. These might be virus particles, or dysfunctional molecules in the human body itself. If you're looking for a drug to do a specific job, a potential strategy is 'rational design'. Through techniques like X-ray crystallography (see page 88), it's possible to glean enough information about a disease molecule to design drug molecules that might interact with it, perhaps blocking it from doing damage in the body. Some of the early work can be carried out in computer simulations, before the candidate drug molecule has even been made in the lab.

Rational design is one strategy that chemists are now employing to deal with one of the biggest problems faced by the pharmaceutical industry today: drug resistance. As microbes and viruses adapt with frightening speed to evade our chemical weapons, the only way to keep them at bay will be to come up with new modes of attack – entirely new classes of drugs. Meanwhile, another frontier of chemistry is to design molecules that could deliver these drugs to specific sites in the body – just one aspect of the new science of nanotechnology.

> **WE [HOPE] THAT ENTERPRISING AND HEARTY ORGANIC CHEMISTS WILL NOT PASS UP THE UNIQUE HEAD START THAT NATURAL PRODUCTS PROVIDE IN THE QUEST FOR NEW AGENTS AND NEW DIRECTIONS IN MEDICINAL DISCOVERY.**
> Rebecca Wilson and Samuel Danishefsky writing in *Accounts of Chemical Research*

The condensed idea
Natural and synthetic routes to disease-beating chemicals

45 Nanotechnology

Only a few decades ago, one of the great scientists of the 20th century came up with some wacky ideas about molecular manipulations and miniature machines. With hindsight, they don't seem half as wacky – they seem like accurate predictions for what nanotechnology has to offer.

The physicist Richard Feynman, one of the scientists involved in developing the atomic bomb and investigating the Challenger Space Shuttle disaster, gave a famous lecture about 'the problem of manipulating and controlling things on a small scale'. This was in 1959 and at that time his ideas seemed so far-fetched they were almost fantastical. He didn't use the term 'nanotechnology' – the word didn't exist until 1974, when a Japanese engineer invented it – but he did talk about moving around individual atoms, building nanomachines that would act as tiny mechanical surgeons and writing an entire encyclopedia on the head of a pin.

A few decades on from Feynman's flight of fantasy, how much of it has become reality? Can we, for instance, manipulate individual atoms? Absolutely – in 1981, the scanning tunnelling microscope was invented granting scientists a first peek into the world of atoms and molecules. Later, in 1989, Don Eigler at IBM realized he could use the probe tip on the machine to push atoms around, spelling out 'IBM' using 35 atoms of xenon. By this time, nanoscientists also had another powerful tool in the form of the atomic

TIMELINE

1875	1959	1986
Discovery of colloidal 'ruby' (nanostructured) gold	Richard Feynman gives his 'There's Plenty of Room at the Bottom' talk	Invention of the atomic force microscope

force microscope and Eric Drexler had written his controversial book about nanotechnology, *Engines of Creation*. Nanotechnology had truly arrived.

REBRANDING SMALL STUFF

Today, thousands of products, from face powders to phones, already contain materials of nano-proportions. The potential applications cover every industry, from healthcare to renewable energy to construction. But nano-stuff is not a human invention. Nano-sized things have been around for longer than we have.

Nanoparticles are exactly what they sound like – very small particles, generally taken as being anything in the range of 1–100 nanometres, or 1–100 millionths of a millimetre. This is at the scale of atoms and molecules – a scale that chemists should be pretty comfortable with, since they spend most of their time thinking about atoms and molecules, and how they behave in chemical reactions. In most substances, atoms clump together in 'bulk' materials, but a gold atom in a big hunk of gold, for example, has radically different properties to a gold nanoparticle, which might contain only a few atoms of the metal. We can turn gold into gold nanoparticles in the lab, but there are plenty of substances that exist in nano-proportions naturally.

The discovery of carbon 'buckyballs' (see page 112) – one-nanometre-wide balls made up of 60 carbon atoms – is often viewed as a landmark in the history of nanoscience, but they are entirely natural. Sure, you can produce buckyballs in the lab, but they also form in the soot of candle flames. Scientists have been making nanoparticles unwittingly for centuries. The 19th-century chemist Michael Faraday did experiments with gold colloids – used to stain glass windows – not knowing that the gold particles were nano-sized. It only became apparent in the 1980s after the arrival of nanotechnology.

> **I AM NOT AFRAID TO CONSIDER THE FINAL QUESTION AS TO WHETHER, ULTIMATELY ... WE CAN ARRANGE THE ATOMS THE WAY WE WANT; THE VERY ATOMS, ALL THE WAY DOWN!**
>
> Richard Feynman (1959)

1986	1989	1991	2012
Eric Drexler publishes *Engines of Creation: The Coming Era of Nanotechnology*	Don Eigler manipulates individual xenon atoms to spell out 'IBM'	Discovery of carbon nanotubes	Announcement of transistor made from a single atom of phosphorus

Nanotube electronics

Nanotubes are tiny tubes of carbon that are incredibly strong and can conduct electricity. They may replace silicon in electronics applications and have been used to make transistors in integrated circuits. In 2013, Stanford University researchers built a simple computer with a processor made from 178 nanotube-containing transistors. It could only run two programs at the same time and had the computing power of Intel's very first microprocessor. One difficulty with using nanotubes in transistors is that they are not perfect semiconducting materials – some form metallic nanotubes that 'leak' current. One US team found that depositing copper oxide nanoparticles on nanotubes helped improve their semiconducting properties.

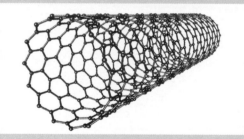

SIZE MATTERS

We can't assume that nanotechnology is nothing new or exciting, however. And we can't pretend that the materials are the same stuff, 'just smaller', because they aren't. Things don't work the same way at the nanoscale as they do in bulk. Perhaps most obviously, smaller particles and materials with nano-sized features have a lot more surface area (per unit volume), which is especially important if you're trying to do chemistry with them. Even weirder than that, things don't look or behave the same. The colour of gold nanoparticles, for example, depends on their size. Faraday's gold colloids weren't golden. They were ruby red.

This weirdness can be useful – gold colloids have been used since antiquity in stained-glass windows – but it may also be problematic. Silver nanoparticles are increasingly being used in antimicrobial dressings, without any great deal of knowledge about how the tiny particles will react in the environment, when they are washed into the water supply. What will be the impact of growing quantities of these particles?

THE REALMS OF FANTASY

Meanwhile, scientists continue to work from the bottom up (see page 100) to create nano-sized objects and devices. An endless realm of possibility stretches ahead, not just of nanoparticles, but of nanomachines, too. So could tiny machines revolutionize medicine, as Feynman imagined? '[It] would be interesting in surgery if you could swallow the surgeon,' he said in his 1959 talk. 'You put the mechanical surgeon inside the blood vessel and it goes into the heart and looks around.' Feynman's nano-surgeon may not be a reality yet, but we can't write it off as fantasy either. Researchers are already working on drug-delivering nanomachines capable of carrying their cargo into diseased cells while avoiding the healthy ones.

Still, we certainly needn't delve into the realms of science fiction to find real-world uses for nanotechnology. Samsung is already incorporating nanostructured materials into the electronic display panels in its phones. Nanotechnology is creating better catalysts for processing fuels and reducing vehicle emissions. Sunscreens have contained titanium dioxide nanoparticles for years – despite recent concerns about their safety.

So, what about writing an entire encyclopedia on a pinhead? No problem. In 1986, Thomas Newman from the California Institute of Technology etched a page of Charles Dickens' *A Tale of Two Cities* into a piece of plastic six thousandths of a millimetre square, making it perfectly feasible to put the *Encyclopedia Britannica* on a two millimetre-wide pinhead.

DNA delivery

Nanoscale building materials can be completely human-engineered or natural. Natural materials have the advantage of being more biocompatible – the body already recognizes them so it is less likely to reject them. That's why some scientists have been working on DNA as an option for delivering drugs. For example, researchers have trapped drug molecules inside DNA cages, with 'locks' that open only with the correct 'keys' – these could be recognition molecules on the surface of cancer cells.

The condensed idea
Small stuff, big impact

46 Graphene

Who knew that a lump of graphite, much like the lead of a pencil, contained a supermaterial so strong, so thin, so flexible and so electrically conductive that it would put any other material on the planet to shame? Who knew that getting it out of there would be so easy? And who knew that doing that might change our mobile phones for ever?

Andre Geim, one of the winners of the 2010 Nobel Prize in Physics, entitled his Nobel lecture 'A Random Walk to Graphene.' By his own admission, he was involved in many unsuccessful projects over the years, with a certain degree of randomness in those he ended up pursuing. Speaking at Stockholm University, Geim said, 'There were two dozen or so experiments over a period of approximately 15 years and, as expected, most of them failed miserably. But there were three hits: levitation, gecko tape and graphene.' Of the three, levitation and gecko tape sound the more interesting, but graphene is the one that's taken the scientific world by storm.

Graphene, often dubbed a 'supermaterial', is the first and most exciting in a new generation of so-called 'nanomaterials – the only substance we know of that consists of a single layer of atoms. Made entirely out of carbon, it is the thinnest and lightest material on the planet but also the strongest. It has been said that a metre-square sheet of graphene – a one-atom-thick layer of carbon, remember – could make a hammock strong and flexible enough to hold a cat, despite weighing about the same as one of the cat's whiskers. A graphene cat hammock would also be transparent, giving the impression of

TIMELINE

1859	1962	1986
Benjamin Brodie discovers 'graphon', which we now know was graphene oxide	Ulrich Hofmann and Hanns-Peter Boehm find very thin fragments of graphene oxide under a transmission electron microscope	Boehm introduces the term 'graphene'

the cat dangling in mid-air, and better than copper at conducting electricity. If you believe the hype, graphene will make it possible to replace batteries with ultrafast-charging 'supercapacitors', putting an end to our phone battery woes and allowing us to charge our electric cars in minutes.

THE FUTURE OF ELECTRONICS

While Geim can't claim to have discovered this supermaterial exactly – other scientists knew of its existence and came pretty close to obtaining it – he and his Nobel co-winner Konstantin Novoselov found a reliable, if not commercially viable, method for producing graphene from graphite. All they did was to get a lump of graphite (see page 112) and use some sticky tape to peel a layer of graphene off its surface. Graphite is the same stuff that's in pencil lead and is basically a stack of hundreds of thousands of graphene sheets with fairly weak attractions between the sheets. With just tape, it's possible to pull off some of the top sheets. Geim and Novoselov didn't realize this until they took a closer look at some tape that had been used to clean a piece of graphite.

> **GRAPHENE HAS LITERALLY BEEN BEFORE OUR EYES AND UNDER OUR NOSES FOR MANY CENTURIES BUT WAS NEVER RECOGNIZED FOR WHAT IT REALLY IS.**
>
> Andre Geim

Although there is some disagreement about exactly who first isolated graphene and when, there's no doubt that the papers the pair published in 2004 and 2005 changed a lot of scientists' minds about the material. Until then, some researchers didn't believe a one-atom-thick sheet of carbon would be stable. The 2005 study went on to probe graphene's extraordinary electronic properties, which have since attracted a lot of attention. There has been plenty of talk of graphene transistors and flexible electronics, including bendy phones and solar cells.

In 2012, two researchers at the University of California, Los Angeles, announced they had made micro-supercapacitors using graphene – akin to very small, long-lasting batteries that charge up in seconds. Graduate

1995	**2004**	**2013**
Thomas Ebbesen and Hidefumi Hiura imagine graphene-based electronics	Andre Geim and Konstantin Novoselov publish a method for getting graphene from graphite	Maher El-Kady and Richard Kaner publish a method for making graphene-based super-capacitors using a DVD burner

Graphene tennis racquets

It's not all about the electronic properties – anything that's three hundred times stronger than steel while weighing less than a milligram a square metre must have other uses too. That's presumably why, in 2013, sports equipment manufacturer HEAD announced it was incorporating graphene into the shaft of its new tennis racquet. That racquet was used by Novak Djokovic when he won the Australian Open later in the year. No one can say whether the win had anything to do with the graphene, but it's a good way to sell tennis racquets.

student Maher El-Kady realized that he could have a lightbulb burn for at least five minutes after charging it for just a couple of seconds with a piece of graphene. He and his supervisor Richard Kaner soon found a way to make their devices using the laser in a DVD burner and are intent on scaling up their production process so that the tiny power sources can be incorporated into everything from microchips to medical implants, such as pacemakers.

GRAPHENE SANDWICH

The fact that graphene is such a good electrical conductor is down to each carbon atom in its flat, chicken-wire-like structure having a free electron. These free electrons shoot across the surface, acting as charge carriers. If there's a problem, it's that graphene is actually too conductive. The semiconductor materials, like silicon (see page 96), that chip manufacturers use to make computer chips are useful because they conduct electricity under certain conditions but not under others – the conductivity can be switched on and off. That's why materials scientists are working on adding impurities to graphene, or even sandwiching it between other superthin materials, to create materials with more tunable electrical properties.

The other problem is producing graphene on a mass scale is not that straightforward, or cheap. Of course, it's not practical to keep peeling it off lumps of graphite. Ideally, materials scientists would also like to have access to larger sheets. One of the more successful methods is chemical vapour deposition, which is a way of sticking gaseous atoms of carbon to a surface to form a layer, but this route requires extremely high temperatures. Other, cheaper methods have been tested, involving industrial-sized kitchen blenders or ultrasound to split layers of graphene from bulk graphite.

Chicken-wire structure

Graphene's structure is often referred to as being like chicken wire. As in graphite, the carbon atoms lie in a single, flat layer, connected by very strong bonds that are difficult to break. Each carbon atom is bonded to three other carbon atoms, forming a repeating pattern of hexagons. This leaves one of the four electrons in each carbon atom's outer shell free to 'wander'. The chicken-wire structure is what gives graphene its strength, while the free electrons give the material its conductivity. A carbon nanotube (see page 180) has a very similar structure – like a piece of chicken wire that has been rolled into a cylinder. Because graphene is one atom thick and completely flat it is considered to be a two-dimensional material, as opposed to three-dimensional, like just about everything else. The fact that it is made entirely of carbon, which is the fourth most common element on Earth, makes it very attractive because it's pretty unlikely we'll ever run out.

DID SOMEONE MENTION LEVITATION?

So that's graphene. What about Geim's other experiments? He levitated water when he, on a whim, poured it into his lab's electromagnet. Geim once even levitated a small frog in a ball of water. The gecko tape was meant to mimic the sticky skin on the foot of a gecko, but it didn't work quite as well as the foot of an actual gecko, so the idea never gathered momentum.

The condensed idea
Supermaterial made from pure carbon

47 3-D printing

Printing might not seem like a subject to get excited about, but that's to ignore the extraordinary possibilities of 3-D printing. From plastic cars to bionic ears made from hydrogels, almost nothing limits the potential of this new technology – aerospace engineers are even printing metal parts for rockets and planes.

In the 20th century, manufacturing was all about mass production. You designed a product that you thought would, on average, suit just about everyone and then you found a way to make that product in vast quantities. Mass production of cars. Mass production of cherry pies. Mass production of computer chips.

So what does the 21st century hold in store? Mass customization – on-demand of consumer products, tailored to suit individual needs and delivered en masse. No longer will we be forced to settle for standard products that suit the 'average person' (no one in particular). You want to adjust the driver's seat in your car to give you a really snug ride without fiddling with levers? Mass customization will allow you to do that. The answer to how manufacturing can adapt to give everyone exactly what they want is 3-D printing.

THE PROMISE OF PRINTING

Printing has for a long time been the domain of chemists. Thousands of years ago, printing inks were made from natural materials and usually contained carbon as a pigment. Today's printing inks are complex mixtures of chemicals including coloured pigments, resins, anti-foaming agents and

TIMELINE

1986	1988	1990
Charles Hull establishes 3-D Systems and is granted patent for stereolithography	First commercial 'Stereolithography Apparatus', the SLA-250, marketed by 3-D Systems	Patent granted to Scott Crump for fused deposition modelling

thickeners. Meanwhile, 3-D printers print with everything from plastic to metal. Some 3-D printers can only print with one type of material – like a black and white printer – while others combine different materials in the same object, as an ordinary printer combines different-coloured inks.

The feature that all 3-D printing techniques have in common is that they build up their structures, layer-by-layer, based on information in a digital file that breaks down three-dimensional objects into two-dimensional cross-sections. Computer-aided design (CAD) programs allow product designers to create complex designs and print them out quickly, rather than assembling them painstakingly from a zillion different parts. The ultimate dream of aerospace engineers is to be able to print a satellite. But some of the structures already created by 3-D printers are truly incredible – bionic ears, skull implants (see 3-D printing body parts, page 191), rocket-engine components and nanomachines, not to mention full-size demonstration cars.

> IMAGINE YOUR PRINTER LIKE A REFRIGERATOR THAT IS FULL OF ALL THE INGREDIENTS YOU MIGHT REQUIRE TO MAKE ANY DISH IN JAMIE OLIVER'S NEW [RECIPE] BOOK.
>
> Lee Cronin

3-D PRINTING INKS

Reliably printing objects like cars and rocket engines is going to require progress in techniques for printing with metals. This is a field that interests the folks at NASA as well as the European Space Agency, which has set up a project called Amaze to print rocket and plane parts. The advantages are a greener, zero-waste production process and the ability to print much more complex metal parts, because they can be built up layer-by-layer.

The 3-D printing process, and the 'ink', depends on the technique. There are already a range of different 3-D printing techniques in development. The process that most closely resembles old-fashioned printing is 3-D inkjet printing, which prints powders and binding materials in alternating layers

1993	2001	2013	2014
MIT researchers are the first to call their device a '3-D printer'	3-D structures printed using inkjet printers	NASA announces it has been testing a 3-D printed rocket engine injector	Patient with bone disorder receives a 3-D printed skull implant

Printing chemicals

A team at the University of Glasgow has been working on adapting 3-D printers to print miniature chemistry sets, into which they can inject the reactants 'inks' for making complex molecules. One potential use of this system could be in making drugs on demand, and cheaply, according to instructions provided by a drug designer's 'software'.

to form a diverse range of materials including plastics and ceramics. Stereolithography, on the other hand, uses a beam of ultraviolet light to activate a resin. The beam draws the design into the resin, layer by layer, causing it to solidify in the shape of the intended structure. In 2014, researchers at the University of California, San Diego, used this approach to print a biocompatible device made from hydrogels that functions like a liver, and can sense and trap toxins in blood.

Perhaps the most widely used 3-D printing technique, though, is fused deposition modelling, which layers up semi-molten materials. Plastics are heated up as they're fed to the printing nozzle straight off the roll. The German engineering company EDAG created the framework for its futuristic-looking 'Genesis' car from thermoplastic, using a modified fused deposition modelling process, and claimed that it would be possible to do the same thing using carbon fibre to build an ultralight, ultrastrong car body. Given that Boeing already makes its Dreamliner aircraft from carbon fibre, why not a 3-D printed plane?

SCALING IT DOWN

From the very big to the very small, 3-D printing is changing the way we design and create. Microfabrication of electronic devices (see page 96) is one very promising area – it's already possible to print electronic circuits and microscale features on lithium ion batteries. Electronics enthusiasts also have the power to quickly design and create customized electronic circuits. Kickstarter funding allowed one company, Cartesian, to develop a printer that would enable the user to print circuits onto different materials, including fabrics, to make wearable electronics.

Nanotechnologists are already surveying the options for printing nanomachines. One technique uses the tip of an atomic force microscope to print molecules onto a surface. However, it's difficult to control the flow of 'ink' at this level. A possible solution is electrospinning, which spins a charged polymer onto an oppositely charged printing surface. Patterns can be incorporated into the surface to control where the materials stick.

It's no wonder that everyone is excited about 3-D printing – the creative possibilities are endless. From the customer's perspective, there are obvious benefits too –no more mass production, a carbon-fibre car with custom-made seats – even perfectly matched replacement body parts.

3-D printing body parts

In September 2014, a paper in the journal *Applied Materials & Interfaces* reported that a team of Australian chemists and engineers had 3-D-printed human-cartilage-mimicking materials. They made them out of high water content hydrogels reinforced with plastic fibres. The two components were printed simultaneously as liquid inks and then 'cured' with UV light to harden them. The result was a tough but flexible composite (see page 168) a lot like cartilage. If you think that's impressive, you obviously haven't heard about the patients who have recently received 3-D-printed skull implants. In 2014, the University Medical Centre Utrecht in the Netherlands announced it had used 3-D printing to replace a large section of skull in a woman with a bone condition that was causing her own skull to thicken, causing brain damage. A Chinese man who lost half of his skull in an accident on a construction site received a new 3-D-printed version made from titanium. The process means it is becoming possible to create custom-designed and fitted implants for every patient.

The condensed idea
Custom creations layer by layer

48 Artificial muscles

How do you get a colossal amount of power out of something that looks pretty flimsy? Think about the skinny cyclists you see powering up French mountains in the Tour de France. It's all about power-to-weight ratio, but how do you do that artificially? The field of artificial muscle research is already producing materials with more impressive stats.

If you've ever fallen into a conversation with a half-decent cyclist, you'll know these guys are nuts for their stats. They're constantly tracking their average speed and totting up their distance and elevation. They share their data on GPS apps and compete over 'KOMs' – 'king of the mountain' records for timed hill climbs. Most of all, they're obsessed with power-to-weight ratios. Any cyclist worth their bicycle clips will know that to win the Tour De France, you've got to have a power-to-weight ratio of about 6.7 watts per kilogram (W/kg).

To the rest of us, that means you've got to be able to pedal like an absolute demon, while being so skinny it looks as if you'd fall off your bike in a sharp gust of wind. Four-time Olympic gold medallist Bradley Wiggins winning the 2011 Tour de France is a case in point. At that time, waiflike Wiggins weighed around 70 kg and could put out about 460 Watts of power. (This may sound impressive, but it would take at least two Bradley Wiggins to power a hairdryer.) This meant he could generate 6.6 Watts of power for every kilogram of his body weight, giving him a power-to-weight ratio of 6.6 W/kg

POWER TO WEIGHT

There's a similar sort of obsession with power-to-weight ratio in the car industry – a 2007 Porsche 911 can put out around 271 W/kg – and also in the scientific field of artificial muscles. For decades, materials scientists have been trying to create materials and devices that can contract like human muscle, but ideally at very high power-to-weight ratios. This opens up the tantalizing possibility of superpowerful robots, which can pull funny faces.

> **ALTHOUGH THE GEL IS COMPLETELY COMPOSED OF SYNTHETIC POLYMER, IT SHOWS AUTONOMOUS MOTION AS IF IT WERE ALIVE.**
>
> Shingo Maeda and colleagues writing in the *International Journal of Molecular Sciences* (2010)

With the current technology, a robot that could lift really heavy weights or, say, cycle up a mountain at speeds approaching the speed of sound, it would need to be pretty bulky to be able to produce enough power. What would be ideal is a robot that doesn't take up too much space but can produce a shedload of power. (And once you have gone to all the effort of creating such a robot and making muscles for it, you might as well use some of them to give your robot a smile or gurn!)

SHRINK AND GROW

The next question, of course, is how does one make tiny, superpowerful muscles? Unsurprisingly, it's not easy. Firstly, a material needs to be found that can expand and contract quickly, like real muscle – it also needs to be stronger than steel without being too stiff. Then you need to find a way to provide energy to that material. The good thing about Bradley Wiggins is that his leg muscles are already packed full of chemical-energy-producing cells, which he supplies with fuel and oxygen just by eating and breathing. However, this exquisite system does not work for a robot.

Most artificial muscles – also called actuators – are polymer-based. In the field of electroactive polymers, scientists are working on soft materials

2011	2012	2014
Bradley Wiggins power-to-weight ratio 6.6 W/kg	Artificial muscles made from nanotube 'yarns'	Polythene muscles have power-to-weight ratio of 5,300 W/kg

Polythene power

Artificial muscles created by chemist Ray Baughman and his team in 2014 were made from four polythene fishing lines twisted together to make a thread 0.8 millimetres thick. Yet upon contraction, this thin thread – made, not from futuristic materials, but from a five-dollar-a-kilo polymer discovered 80 years previously – was capable of lifting a weight equivalent to a medium-sized dog and of contracting by half its length. How can a barely visible bundle of fishing lines lift a 7-kilogram load? The answer lies in the twisting and then coiling of the polythene, which turns it into a torsional material and allows it to withstand much greater strains. Many artificial muscles take their energy from electricity, but the polythene threads respond to simple changes in temperature. To get them to contract, you apply heat and as they cool down, they relax. The 'muscles' can be encased in tubes so that they can be quickly cooled using water. The only problem is changing the temperature quickly enough to replicate ultrafast muscle-twitching.

that change shape and size when they are connected to an electric current. Silicone and acrylic materials known as elastomers make good actuators and some are already commercially available. There are also ionic polymer gels that swell or shrink in response to an electrical current or a change in chemical conditions. Any artificial muscle needs a source of energy, but those materials that depend on electricity sometimes need a constant power supply to keep them contracting.

In 2009, however, Japanese researchers made a piece of polymer gel 'walk' unaided, using nothing but chemistry – a classic chemical reaction called the Belousov-Zhabotinsky reaction. In this reaction, the quantity of ruthenium bipyride ions constantly oscillates, affecting the polymers by making them shrink and grow. In a curved strip of gel, this translates into autonomous movement – as the researchers themselves wrote, 'as if it were alive'. Like a caterpillar slowly inching its way across a surface, it wasn't very quick but it was utterly mesmerizing to watch.

DO THE TWIST

More advanced – and far more expensive – materials have been made using carbon nanotubes (see page 180). In the last few years, these materials have started to approach pinnacles of superstrength, superspeed and superlightness that would, quite frankly, put Wiggins to shame. In 2012, an international team including researchers at the NanoTech Institute at the University of Texas at Dallas announced they had made artificial muscles using carbon nanotubes that had been twisted to make yarn and then filled with wax. These nanotube yarns could lift 100,000 times their own weight, contracting in 25 thousandths of a second when connected to a current. Those mindblowing performance figures for a wax-filled yarn give a phenomenal power-to-weight ratio of 4,200 W/kg. That's several orders of magnitude above the power density of human muscle tissue.

Nanotubes are some of the strongest materials known to humankind, but at several thousand dollars per kilo, they're also very pricey. Convinced they could do it on a tighter budget, the researchers went back to the drawing board. Two years later, they announced they had repeated the feat using coiled polythene fishing lines (see Polythene power, opposite). The cheap artificial muscles they had made absorbed their energy from heat and were capable of lifting a 7.2 kg weight despite being less than a millimetre thick. The power density of this Heath Robinson device was an incredible 5,300 W/kg. Take that, Bradley Wiggins!

Not just for robots

Apart from facial expressions for robots (and lifting heavy weights), what else can you use artificial muscles for? Other ideas include human exoskeletons, precise control of microsurgery, positioning of solar cells, and also clothes with pores that shrink and expand depending on the weather. Using woven polymer muscles that contract or relax in response to temperature changes, it could be possible to create fabrics that literally breathe. Similar concepts are behind designs for self-opening shutters or blinds

The condensed idea
Materials that make like real muscles

49 Synthetic biology

Advances in chemical synthesis of DNA mean that scientists can now patch together genomes of their own design to create organisms that don't exist in nature. Sounds a bit ambitious, doesn't it? But building synthetic organisms from the bottom up might one day be as simple as snapping together building bricks.

Synthetic biologists don't follow recipes. But instead of improvising in the kitchen, as you might when cooking a chilli con carne, they improvise in the lab with life itself. While their creations have so far remained faithful to nature's recipe book, they have ambitious plans. In the future, they plan to create the synthetic-biological equivalent of a chilli con carne made with crocodile meat and edamame beans – not what you or I would recognize as chilli.

> **WE'RE GOING TO BE ABLE TO WRITE DNA. WHAT DO WE WANT TO SAY?**
> Synthetic biologist Drew Endy

REINVENTING NATURE

The fledgling field of synthetic biology has grown out of biologists' desire to improve on nature by editing the genomes of living organisms. It all began with genetic engineering – a technique that has proved really useful in animal studies seeking to understand the role of certain genes in disease. Now, alongside advances in DNA sequencing and synthesis, this has progressed to projects covering whole genomes.

Whereas traditional genetic engineering might change a single gene to study the effect that it would have on an animal, a plant or a bacterium, synthetic biology might edit out thousands of 'letters' (bases) of DNA code

TIMELINE

1983	1996	2003	2004
PCR – fast new chemical process for synthesizing DNA developed	Yeast genome mapped	Registry of Standard Biological Parts founded	First international synthetic biology meeting held at MIT

and introduce genes encoding entire metabolic pathways for molecules that an organism has never produced before. One of the first projects hailed as a triumph for synthetic biology was the re-engineering of yeast to produce a chemical precursor for the antimalarial drug, artemisinin. The French pharmaceutical company Sanofi finally launched production of its semi-synthetic version of the drug in 2013, with the aim of making up to 150 million treatments in 2014. Even so, some scientists preferred to see this as a sophisticated genetic-engineering project involving just a handful of genes – impressive but far from a 'crocodile and edamame-level' redesign.

Making DNA from scratch

One of the advances in DNA synthesis that drove big cost reductions was the development of a synthetic chemistry process that uses molecules called phosphoramidite monomers. Each monomer is a nucleotide (see DNA, page 140) like that in ordinary DNA, except that it has caps over its reactive bits. These chemical caps are only removed (deprotection), using acid, just before new nucleotides are added to growing DNA chains. The first nucleotide, carrying the correct base (A, T, C or G) is anchored to a glass bead. New nucleotides are then added in cycles of deprotection and coupling, in the order that creates the desired code. In most cases, only short stretches are synthesized. Many short pieces are then stitched together. Of course, in the case of the synthetic biologist, the code might not belong to any

natural organism – it can be completely of their own design. Phosphoramidite chemistry currently dominates the DNA synthesis industry and it is expected that really significant reductions in the speed and cost of synthesis will now require another new type of chemistry. Other chemical routes are possible but none has become commercial just yet.

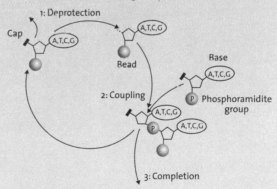

Dangerous puzzle

In 2006, reporters at *The Guardian* newspaper managed to buy smallpox DNA online. The vial they received in the post only contained a segment of the smallpox genome, but the newspaper claimed that a well-funded terrorist organization would only need to 'order consecutive lengths of DNA along the sequence and glue them together' to create a deadly virus. DNA synthesis companies now screen orders for dangerous sequences but some scientists argue samples of such potentially devastating killers should be destroyed.

MAIL-ORDER DNA

Meanwhile, Craig Venter, the geneticist famous for his involvement in sequencing the human genome, had been working on a completely synthetic genome. In 2010, his team at the J. Craig Venter Institute announced it had pieced together the genome – with a few minor modifications – of the goat parasite *Mycoplasma mycoides* and stuck it inside a living cell. While Venter's synthetic genome was basically a copy of the real thing, it did demonstrate the creation of life using exclusively synthetically made DNA.

All this only became possible because of advances in 'reading and writing' DNA that allowed researchers to sequence and chemically synthesize DNA sequences (see Making DNA from Scratch, page 197) rapidly and relatively cheaply. During the years that Venter and his competitors were unravelling the human genome (1984 to 2003), the cost of both sequencing and synthesizing DNA dropped dramatically. By some estimates, you can now get an entire human genome of over three million base pair sequenced for $1,000 and it costs just 10 cents per base to make DNA.

These price cuts have given synthetic biologists access to the instructions for making many organisms that they might like to re-engineer, or steal from, and allow them to test out their designs for new ones. They don't even have to make the DNA themselves. They can just send their sequences off to a specialist synthesis company and have the DNA mailed back to them in the post. It sounds like cheating but to return to the chilli con carne analogy, it's equivalent to buying a pre-made spice mix to make your Mexican masterpiece, instead of going to all the trouble of chopping up fresh chillies and grinding cumin seeds.

STANDARD BIOLOGICAL PARTS

Another way synthetic biologists plan to cut their workload is by building a database of standard parts that can be used to assemble synthetic organisms. This has already been in development since 2003, in the form of the Registry of Standard Biological Parts. Less gruesome than it sounds, the registry is a collection of thousands of 'user-tested' genetic sequences shared by the synthetic biology community. The idea is to make compatible parts with known functions that snap together like building bricks to construct working organisms from scratch. One of these bricks might code for a colourful pigment, for example, while another might code for a genetic master switch that activates a whole series of enzymes when a specific chemical is sensed.

The ultimate goal of synthetic biology is to be able to piece together the genomes of human-designed organisms capable of producing novel drugs, biofuels, food ingredients and other useful chemicals. Before we get ahead of ourselves, though, it's worth noting that we're a long way from being able to make, say, synthetic crocodiles for our crocodile chilli. As far as more complex organisms go, the furthest we've got is fungi.

Though you might not think of brewer's yeast as being particularly advanced, on a cellular level we've got more in common with yeast than with bacteria. The Sc? n project aims to build a redesigned, synthetic version of the yeast *Saccharomyces cerevisiae* (see page 56), chromosome-by-chromosome. Operating a 'take stuff out till it breaks' approach, the international team has been trying to streamline the genome by removing all non-essential genes, and then inserting small pieces of their synthetic code into natural yeast to check it still works. They've only finished one chromosome so far. The results could be ruinous (for the yeast, anyway) or they could be a revelation, but the team hope to discover exactly what it takes to make a living organism.

The condensed idea
Redesigning life

50 Future fuels

What happens when fossil fuels run out? Are we going to have to power everything with solar panels and wind turbines? Not necessarily. Chemists are working on new ways to make fuels that won't pump carbon dioxide into the atmosphere. The hard bit will be making them without using up even more of Earth's most precious resources.

Two of the biggest technological challenges the world faces today are related to fuels. One: fossil fuels are running out. Two: burning fossil fuels is filling the atmosphere with greenhouse gases, changing the very nature of our planet for the worse. The solution seems blindingly obvious – stop using fossil fuels.

Reducing our reliance on fossil fuels means finding another way to power the planet. While solar and wind power can make big contributions to our energy needs, they're not fuels – you can feed the energy into the national grid but you can't pump it into your car and drive off with it. That's where fossil fuels have the advantage: energy is stored in liquid, chemical form.

But surely electric vehicles have solved this problem already? Why can't we just charge them up using solar power from the grid? Currently, fossil fuels are a much more efficient way of carrying energy around. You can cram

TIMELINE

1800	1842	1820s
Electrolysis of water to produce hydrogen and water	Matthias Schleiden proposes that photosynthesis splits water	Fischer–Tropsch process developed for making fuels from hydrogen and carbon monoxide

Artificial leaves

Artificial leaves or 'water-splitters' tend to be based on a general schema that deals with each half of the water-splitting reaction separately. On each side is an electrode and the two sides are separated by a thin membrane that stops most molecules moving across it. The electrodes on both sides are made from a semiconducting material which, like the silicon in a solar cell, absorbs the energy in light. On one side, the catalyst coated onto the electrode pulls the oxygen from water and on the other side, another catalyst generates the all-important hydrogen by uniting hydrogen ions with electrons. Some devices have used rare, expensive metals like platinum as catalysts, but the search is on for cheaper materials that would be more sustainable to use in the long run. High-throughput approaches that screen millions of potential catalysts are being employed to try to find the best materials. Chemists must consider not only their catalytic abilities, but also their durability, cost and the availability of materials needed to make them. Some researchers are even modelling their catalysts on organic molecules used by plants in real photosynthesis.

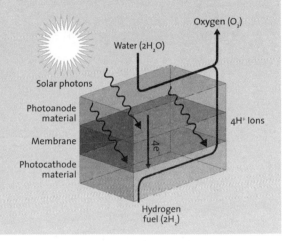

more energy per unit weight into petroleum products, which makes them a pretty unassailable power source for vehicles, such as aeroplanes. Unless there are some significant advances and drastic weight reductions in battery technology, we can build all the solar-power plants and wind turbines we want, but we're still going to need fuels. Furthermore, our energy systems are already fuel-based, meaning that if we could develop cleanly produced alternatives, they might not require such an overhaul.

1998

Unstable artificial leaf created by scientists at National Renewable Energy Laboratory

2011

Low-power artificial leaf announced, costing less than US $50 to produce

2014

Solar-Jet project demonstrates process for making jet fuels with carbon dioxide, water and light

HYDROGEN HEADACHE

One potential solution may lie in the smallest, simplest element at the top of the Periodic Table: hydrogen. Already used as rocket fuel, it seems like the perfect solution. In a hydrogen-powered car, hydrogen would react with oxygen inside a fuel cell to release energy and produce water. It's clean and without a carbon atom in sight, but where do you get an endless supply of hydrogen and how do you carry it around safely? It only takes a little bit of oxygen and a spark to make a really big explosion.

> **RESTORE HUMAN LEGS AS A MEANS OF TRAVEL. PEDESTRIANS RELY ON FOOD FOR FUEL AND NEED NO SPECIAL PARKING FACILITIES.**
> Historian and philosopher,
> Lewis Mumford

Chemists' first challenge is to find an endless source of hydrogen. William Nicholson and Anthony Carlisle made hydrogen in 1800 by sticking the wires of a primitive battery into a tube of water (see page 92). In fact, this 'splitting' of water is what plants do in photosynthesis. As they often do, chemists are trying to copy from nature and are trying to make artificial leaves (see Artificial leaves, page 201).

Artificial photosynthesis has become an epic science project, with governments devoting hundreds of millions of dollars to try to create a workable water-splitter. It's primarily a hunt for materials that harvest sunlight (as in a solar panel) and materials that catalyse the production of hydrogen and oxygen. The focus is now on finding common materials that don't cost the earth, or degrade after just a few days.

OLD PROBLEM, NEW SOLUTION

Assuming we'll be able to do it practically, we could even use the hydrogen to make more traditional fuels. In the Fischer–Tropsch process, a mixture of hydrogen and carbon monoxide (CO), otherwise known as syngas, is used to make hydrocarbon fuels (see page 64). This would do away with the idea of having to create a whole new infrastructure of hydrogen-fuelling stations.

But you can make syngas another way too: heat carbon dioxide and water to 2,200 °C and they break up into hydrogen, carbon monoxide and oxygen.

There a couple of problems with this approach: first, it takes a lot of energy to reach such high temperatures; and second, the oxygen is a serious explosion risk if it gets anywhere near hydrogen. Some of the latest practical water-splitting devices face the same problem, because they don't separate the oxygen and hydrogen that are produced in the water-splitting reactions.

In 2014, however, chemists working on the European Solar-Jet project did something impressive. They turned syngas into jet fuel, via the Fischer–Tropsch process. Although they only made a tiny amount, symbolically it represented a milestone, because they did it using a 'solar simulator' – something simulating a solar concentrator. Solar concentrators are giant, curved mirrors that focus light on a single spot to generate very high temperatures. The researchers used this solar-derived heat to create syngas, thus overcoming the energy problem, and an oxygen-absorbing material – cerium oxide – to deal with the explosion risk.

So in one sense, chemists have solved the problem. They can already make clean fuels, and even jet fuel, using the endless supply of energy in sunlight. It won't be plain sailing from here though. The difficult bit, as so often is the case, will be doing it cheaply, reliably and without using up all of the world's natural resources in the process. Today, clever chemistry isn't just about making what you need; it's about doing it in a way that means you can keep doing it forever.

Hydrogen slaves

One idea for producing hydrogen is to harness green algae that photosynthesize, or plants, to make it for us. Some algae split water, producing oxygen, hydrogen ions and electrons, and then use enzymes called hydrogenases to glue the hydrogen ions and electrons together, producing hydrogen gas (H_2). It might be possible to reroute some of the reactions in these algae, by genetic engineering, to make them produce more hydrogen. Scientists have already identified some of the important genes.

The condensed idea
Clean, transportable energy

The periodic table

Elements in the periodic table are arranged in order of increasing atomic number, and also by recurring trends in their chemical properties. They naturally fall into vertical columns that share similar chemical properties, and horizontal rows (periods) with generally increasing mass.

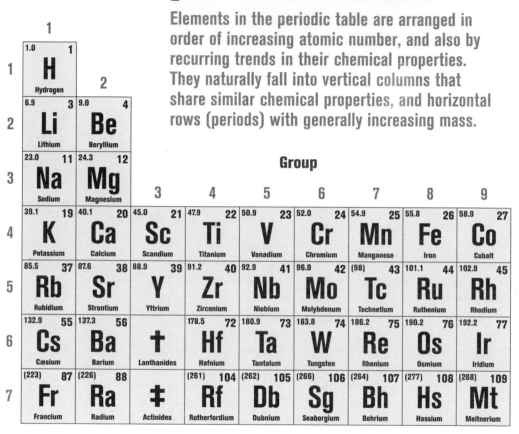

Group

Period

1	2	3	4	5	6	7	8	9
1.0 1 **H** Hydrogen								
6.9 3 **Li** Lithium	**9.0** 4 **Be** Beryllium							
23.0 11 **Na** Sodium	**24.3** 12 **Mg** Magnesium							
39.1 19 **K** Potassium	**40.1** 20 **Ca** Calcium	**45.0** 21 **Sc** Scandium	**47.9** 22 **Ti** Titanium	**50.9** 23 **V** Vanadium	**52.0** 24 **Cr** Chromium	**54.9** 25 **Mn** Manganese	**55.8** 26 **Fe** Iron	**58.9** 27 **Co** Cobalt
85.5 37 **Rb** Rubidium	**87.6** 38 **Sr** Strontium	**88.9** 39 **Y** Yttrium	**91.2** 40 **Zr** Zirconium	**92.9** 41 **Nb** Niobium	**96.0** 42 **Mo** Molybdenum	**(98)** 43 **Tc** Technetium	**101.1** 44 **Ru** Ruthenium	**102.9** 45 **Rh** Rhodium
132.9 55 **Cs** Cæsium	**137.3** 56 **Ba** Barium	**†** Lanthanides	**178.5** 72 **Hf** Hafnium	**180.9** 73 **Ta** Tantalum	**183.8** 74 **W** Tungsten	**186.2** 75 **Re** Rhenium	**190.2** 76 **Os** Osmium	**192.2** 77 **Ir** Iridium
(223) 87 **Fr** Francium	**(226)** 88 **Ra** Radium	**‡** Actinides	**(261)** 104 **Rf** Rutherfordium	**(262)** 105 **Db** Dubnium	**(266)** 106 **Sg** Seaborgium	**(264)** 107 **Bh** Bohrium	**(277)** 108 **Hs** Hassium	**(268)** 109 **Mt** Meitnerium

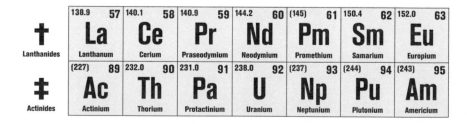

† Lanthanides	**138.9** 57 **La** Lanthanum	**140.1** 58 **Ce** Cerium	**140.9** 59 **Pr** Praseodymium	**144.2** 60 **Nd** Neodymium	**(145)** 61 **Pm** Promethium	**150.4** 62 **Sm** Samarium	**152.0** 63 **Eu** Europium
‡ Actinides	**(227)** 89 **Ac** Actinium	**232.0** 90 **Th** Thorium	**231.0** 91 **Pa** Protactinium	**238.0** 92 **U** Uranium	**(237)** 93 **Np** Neptunium	**(244)** 94 **Pu** Plutonium	**(243)** 95 **Am** Americium

Sample entry:
cobalt

Mass number (average of various isotopes) — 58.9

Atomic number — 27

58.9	27
Co	
Cobalt	

Symbol — Co

Element name — Cobalt

Group

Group			Group	13	14	15	16	17	18
									4.0 2 **He** Helium
10	11	12		10.8 5 **B** Boron	12.0 6 **C** Carbon	14.0 7 **N** Nitrogen	16.0 8 **O** Oxygen	19.0 9 **F** Fluorine	20.2 10 **Ne** Neon
				27.0 13 **Al** Aluminium	28.1 14 **Si** Silicon	31.0 15 **P** Phosphorus	32.1 16 **S** Sulfur	35.5 17 **Cl** Chlorine	39.9 18 **Ar** Argon
58.7 28 **Ni** Nickel	63.5 29 **Cu** Copper	65.4 30 **Zn** Zinc	69.7 31 **Ga** Gallium	72.6 32 **Ge** Germanium	74.9 33 **As** Arsenic	79.0 34 **Se** Selenium	80.0 35 **Br** Bromine	83.8 36 **Kr** Krypton
106.4 46 **Pd** Palladium	107.9 47 **Ag** Silver	112.4 48 **Cd** Cadmium	114.8 49 **In** Indium	118.7 50 **Sn** Tin	121.8 51 **Sb** Antimony	127.6 52 **Te** Tellurium	126.9 53 **I** Iodine	131.3 54 **Xe** Xenon
195.1 78 **Pt** Platinum	197.0 79 **Au** Gold	200.6 80 **Hg** Mercury	204.4 81 **Tl** Thallium	207.2 82 **Pb** Lead	209.0 83 **Bi** Bismuth	(210) 84 **Po** Polonium	(210) 85 **At** Astatine	(220) 86 **Rn** Radon
(271) 110 **Ds** Darmstadium	(272) 111 **Rg** Roentgenium	(285) 112 **Cn** Copernicium	(284) 113 **Uut** Ununtrium	(289) 114 **Fl** Flerovium	(288) 115 **Uup** Ununpentium	(292) 116 **Lv** Livermorium	(294) 117 **Uus** Ununseptium	(294) 118 **Uuo** Ununoctium

157.3 64 **Gd** Gadolinium	158.9 65 **Tb** Terbium	162.5 66 **Dy** Dysprosium	164.9 67 **Ho** Holmium	167.3 68 **Er** Erbium	168.9 69 **Tm** Thulium	173.0 70 **Yb** Ytterbium	175.0 71 **Lu** Lutetium
(247) 96 **Cm** Curium	(247) 97 **Bk** Berkelium	(251) 98 **Cf** Californium	(252) 99 **Es** Einsteinium	(257) 100 **Fm** Fermium	(258) 101 **Md** Mendelevium	(259) 102 **No** Nobelium	(262) 103 **Lr** Lawrencium

Index

First published in Great Britain in 2015 by

Quercus
Carmelite House
50 Victoria Embankment
London EC4Y 0DZ

An Hachette UK company

First published in 2015

Design and editorial by Pikaia Imaging

A CIP catalogue record for this book is available from the British Library

HB ISBN 9781848666672
EBOOK ISBN 9781848666689

10 9 8 7 6 5 4 3 2 1

Printed and bound in China

Acknowledgements:

Huge thanks to all the members of the Chemistry Super-Panel for their thoughts and advice in the development of this book: Raychelle Burks (@DrRubidium), Declan Fleming (@declanfleming), Suze Kundu (@FunSizeSuze) and David Lindsay (@DavidMLindsay). Staff at Chemistry World magazine also provided invaluable help and support – thanks to Phillip Broadwith (@broadwithp), Ben Valsler (@BenValsler), and Patrick Walter (@vinceonoir). Special thanks to Liz Bell (@liznewtonbell) for sanity checks and spreadsheet hilarity in the last two weeks, and as always to Jonny Bennett for doing all the feeding and watering, not to mention everything else. Finally, thank you to James Wills and Kerry Enzor for their understanding through some difficult days at the start of this project, and to Richard Green, Giles Sparrow and Dan Green for guiding it to the end.

Picture credits:

109: Emw2012 via Wikimedia; 191: University of Hasselt; 194: NASA.
All other pictures by Tim Brown.